SCIENCE SCOPE
PHYSICS

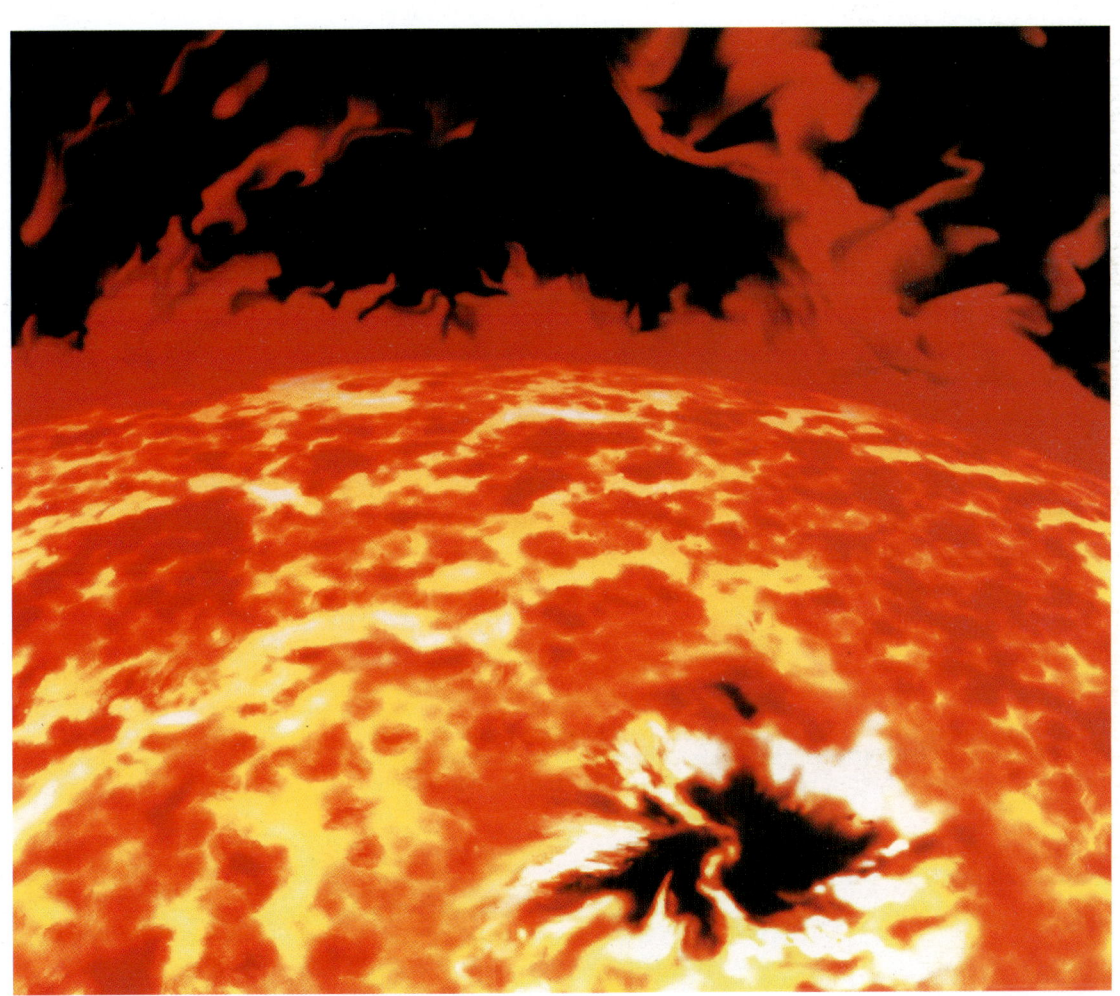

Brian Arnold

Hodder & Stoughton

A MEMBER OF THE HODDER HEADLINE GROUP

Photo Acknowledgements

The publisher would like to thank the following individuals, institutions and companies for permission to reproduce photographs in this book. Every effort has been made to trace ownership of copyright. The publishers would be happy to make arrangements with any copyright holder whom it has not been possible to contact.

Action Plus (25, 134, 148, 160); Andrew Lambert (69 right, 92 left and right); Bill Butland ARPS (72); Bruce Coleman Picture Library (105); Corbis (1, 158 top left); Hodder & Stoughton (8, 21, 115, 118); Life File (2, 9, 13 top, 13 bottom, 18, 56, 63 middle and bottom, 79, 86, 89, 135, 151 bottom left and right); Science Photo Library (42, 54, 61, 63 top, 64, 69 left and middle, 90, 107, 109, 119, 126 left and right, 127, 128, 129, 130, 136 top and bottom, 151 top, 152, 158 top right and bottom).

The publishers would also like to thank the British Library for permission to use the pictures of the heliocentric and geocentric universe (pages 122 and 123) from *A perfit description of Caelestiall Orbes 718.g.52* and *Cosmographia 1007.g.24*, respectively.

Orders: please contact Bookpoint Ltd, 130 Milton Park, Abingdon, Oxon OX14 4SB. Telephone: (44) 01235 827720, Fax: (44) 01235 400454. Lines are open from 9.00–6.00, Monday to Saturday, with a 24 hour message answering service. You can also order through our website www.hodderheadline.co.uk

British Library Cataloguing in Publication Data
A catalogue record for this title is available from The British Library

ISBN 0 340 80478 5

First published 2002
Impression number   10 9 8 7 6 5 4
Year   2008 2007 2006 2005 2004

Copyright © 2002 Brian Arnold

Cover photo from Science Photo Library.
Typeset by Fakenham Photosetting Limited, Fakenham, Norfolk.
Printed in Italy for Hodder & Stoughton Educational, a division of Hodder Headline Ltd, 338 Euston Road, London NW1 3BH.

Contents

Preface

Physics is an exciting subject. It has an impact on all our lives; ranging from the mobile phones we use daily to the nuclear power stations that generate the electricity we use in our homes. I have written this book in the hope that pupils will recognise and understand many of the basic principles on which developments such as these have been based.

This book provides extensive cover for pupils in Years 7, 8 and 9 following the Physical Processes section of the Key Stage 3 Science National Curriculum or the Common Entrance Examination at 13+ Science Syllabus (Physics).

Particular attention has been paid to the inclusion of extension material, which will aid pupils of average and above average abilities, who are aiming for a high level of acheivement.

The following features have been included in the book:

- Test Yourself Questions to consolidate and reinforce understanding.
- Extension Boxes, which contain material aimed specifically at those pupils aiming for the higher tiers or following the Common Entrance examination.
- Summaries bring together all the ideas in the chapter.
- End-of-Chapter Questions provide opportunities to apply knowledge from the topic. Questions that refer to material covered in the extension boxes are on a yellow tinted background.

I hope you have as much fun reading this book as I have had in writing it.

Acknowledgements

My thanks to all at Hodder and Stoughton for their help in producing this book and in particular to Charlotte Litt for seeing the book through all its stages. Finally a big thank you to my wife Jill for her invaluable comments, support and enthusiasm for the project.

Brian Arnold

Matching Grid

	National Curriculum Programme of Study	Key Stage 3 Scheme of Work	Common Entrance Examination 13 +
Chapter 1 Energy sources and changes	5a, 5b, 5c	7I, 9I	5. a, b, c, e, g
Chapter 2 Forces and their effects	2a, 2b, 2c, 2d	7K, 9J	2. b, c, d
Chapter 3 Electrical circuits	1a, 1b, 1c,	7J, 9I	1. a, b, c
Chapter 4 Magnetism	1d,	8J	1. d
Chapter 5 Electromagnetism	1e, 1f	8J	1. e, f
Chapter 6 Heating and cooling		7G, 8I	
Chapter 7 Temperature and the movement of energy	5d, 5e, 5f	8I	5. d
Chapter 8 Rays of light and reflection	3a, 3b, 3c	8K	3. a, b, c
Chapter 9 Refraction of light	3d, 3e, 3f	8K	3. d, e, f
Chapter 10 Sound and hearing	3g, 3h, 3i, 3j, 3k	8L	3. g, h, i, j, k
Chapter 11 The Earth in space	4a, 4b, 4c	7L, 9J	4. a, b, c
Chapter 12 The stars and the Universe	4c, 4d, 4e	7L, 9J	4. c, d, e
Chapter 13 Pressure	2g	9L	2. g
Chapter 14 Moments	2e, 2f	9L	2. e, f
Chapter 15 Speeding up	2a, 2c, 2d	9K	2. a, c, d

1 Energy

D id you know that a roller coaster like this is an 'energy changer'? The designers of the track do. They know all about energy and how it can be used to create a ride that will excite people. When you have finished studying this chapter, you too will understand more about energy and the uses we can put it to.

Types of energy

There are many different kinds of energy:

- We obtain *light energy* from the Sun, stars, fires and light bulbs.

- We obtain *heat energy* from objects that are hot, such as the Sun, fires and heaters.

- We obtain *sound energy* from objects that are vibrating, such as the strings of a guitar or the vocal cords in your throat.

- Food is a source of *chemical energy* which we use to keep us alive. We need energy when we run, walk, play and even when we sleep. Cells and batteries also contain chemical energy.

- **Kinetic energy** or movement energy is the energy an object has because it is moving.

- We obtain *electrical energy* whenever a current flows.

- Objects that are stretched or twisted out of shape have **elastic** or **strain potential energy**.

Figure 1 ▲   The strain energy stored in this bow is going to be used to shoot the arrow. Once the arrow has been released and is flying through the air, it has kinetic energy because of its movement

- Objects that are 'up high' have **gravitational potential energy** or position energy. The higher they are, the more gravitational potential energy they have.

- We obtain *nuclear energy* from reactions that take place in the centres of atoms. These are called **nuclear reactions**.

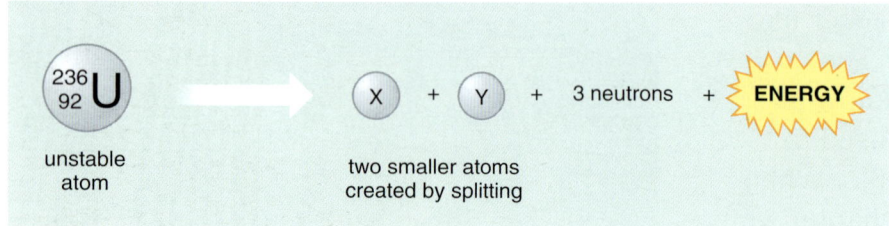

Figure 2 ▲ The water at the top of Niagara Falls in Canada has a lot of gravitational potential energy because it is so high

Figure 3 ▲ When some atoms 'split', they release lots of energy

Test Yourself

1 Name seven different types of energy.

2 Give one source of each of your answers to Question 1.

Energy transformations

Whenever we use energy, it changes into other forms of energy, i.e. it is transformed.

Table 1 lists some examples of **energy transformations**.

Energy at the start	Energy changer	Energy after
electrical	loudspeaker	sound
electrical	light bulb	heat and light
chemical	wood fire	heat and light
light	green plant	chemical
sound	microphone	electrical
chemical	car engine	heat, kinetic and sound
electrical	hair drier	heat, kinetic and sound
strain potential	catapult	kinetic
kinetic	generator or dynamo	electrical

Table 1 ▲ Some energy transformations

Concentrated and dilute forms of energy

Although energy is never lost during an energy transformation, it often changes into a less concentrated form which is not so easily reused. We say that the energy has **dissipated**. For example, the wax of a candle is a **concentrated** form of energy but the light and heat into which it changes are much less concentrated and are therefore difficult to reuse. The light and heat are more **dilute forms of energy**.

We can show the amounts of energies in a simple energy transfer in the form of a diagram. Energy is measured in joules (J) or kilojoules (kJ). One kJ is equal to 1000 J.

(a)

200 J of chemical energy

190 J of heat energy

10 J of light energy

(b)

100 J of electrical energy

75 J of heat energy

20 J of kinetic energy
5 J of sound energy

(c)

50 J of strain energy

25 J of kinetic energy

15 J of heat energy

10 J of sound energy

Figure 4 ▲ Energy transfer diagrams for a) a candle, b) a hair drier and c) a car

Test Yourself

3 What happens to energy when it is transformed?

4 What type(s) of energy
   a) does water have at the top of a waterfall?
   b) is contained in a battery?
   c) is released when a firework is lit?

5 What device would you use to change
   a) kinetic energy into electrical energy?
   b) electrical energy into kinetic energy?
   c) sound into electrical energy?

Test Yourself

6 Explain the difference between a concentrated form of energy and a dilute form of energy.

7 Draw an energy transfer diagram for a radio that changes 200 J of electrical energy into 100 J of sound energy and 100 J of heat.

Energy from food

We obtain the energy we need to live and grow from the food we eat. Our age, sex and lifestyle determine the types of food we eat and how much of each we ought to eat each day.

A man who does physical work needs lots of energy. Each day he will need about 15 000 kJ of energy from the food he eats. A young child will need much less, approximately 8000 kJ a day.

Figure 5 shows some approximate energy needs for one day. Boys and men tend to need a little more energy than girls and women.

Different types of food contain different amounts of energy. Table 2 shows how much energy is contained in an average portion of different foods.

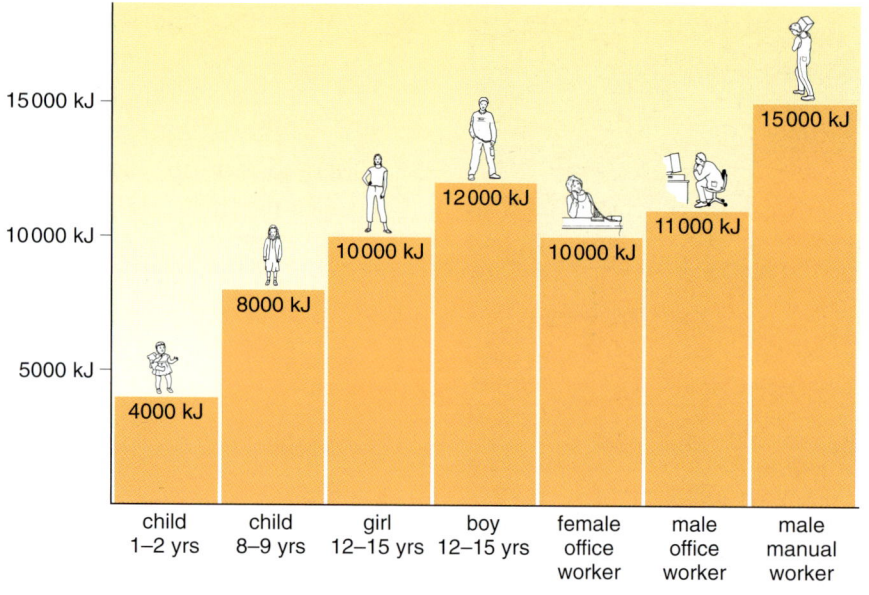

Figure 5 ◄ The approximate daily energy needs of different types of people

Food	Energy content in kJ	Food	Energy content in kJ
apple/orange	200	cabbage	80
banana	300	carrots	80
boiled potatoes	400	lettuce	40
chips	1000	bread and butter	400/slice
boiled rice	500	ice cream	500
spaghetti	500	crisps	600
pizza	1200	cake	700
peas	300	chocolate	1500
baked beans	300	lemonade	700

Table 2 ▲ Energy contents of different foods

Test Yourself

8 From where do we get the energy we need to live and grow?

9 Give two reasons why different groups of people have different daily energy requirements.

10 Why do fell walkers and mountaineers often carry chocolate with them?

Extension box

Efficiency

Efficiency of a light bulb

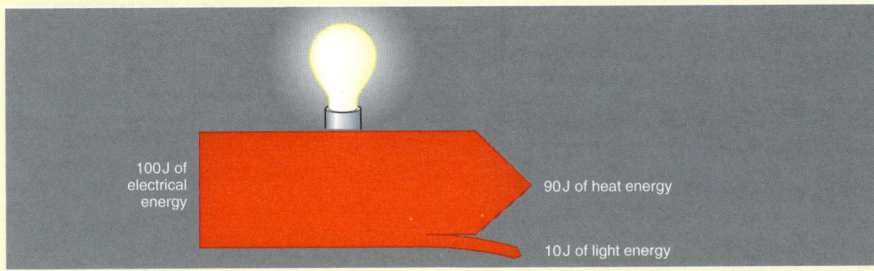

100 J of electrical energy

90 J of heat energy

10 J of light energy

Figure 6 ◄ This light bulb is only 10% efficient

We use light bulbs to change electrical energy into light energy. However, as we have already seen, some of this electrical energy is changed into heat. In fact in the example shown above, only 10 J of every 100 J that enters the light bulb is changed into light. The other 90 J is changed into unwanted heat.

We describe how well a device changes energy into the forms we want using the term **efficiency**. We calculate a value for efficiency using the equation:

efficiency = useful energy out / total energy in × 100%

In the above example, the efficiency of the bulb = 10 J / 100 J × 100% = 10%

Efficiency of a car

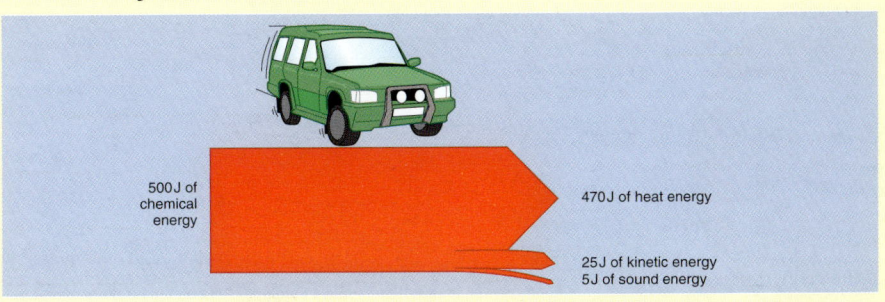

500 J of chemical energy

470 J of heat energy

25 J of kinetic energy
5 J of sound energy

Figure 7 ◄ This car is 5% efficient

Efficiency of a car = 25 J / 500 J × 100% = 5%

Extension box

Efficiency of an electric fire

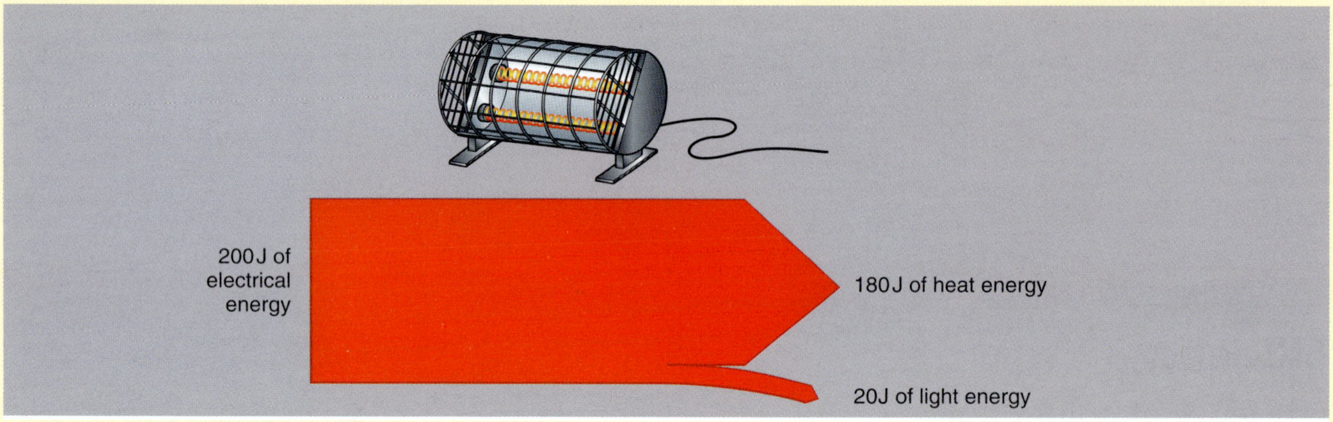

200 J of electrical energy

180 J of heat energy

20 J of light energy

Efficiency of electric fire = 180 J / 200 J × 100% = 90%

Figure 8 ▲ This electric fire is 90% efficient

Test Yourself

11 Calculate the efficiency of an electric train which changes 1000 J of electrical energy into 550 J of heat energy, 50 J of sound energy and 400 J of kinetic energy.

Energy resources

It is important that we use sources of concentrated energy wisely. In the UK, the **fossil fuels** coal, oil and gas provide almost 80% of our energy needs. If we continue to use them at this rate, they will soon be gone. Coal, oil and gas are examples of **non-renewable** sources of energy. Once they have been used, they cannot be replaced.

Figure 9 shows how different energy resources were used in 2000.

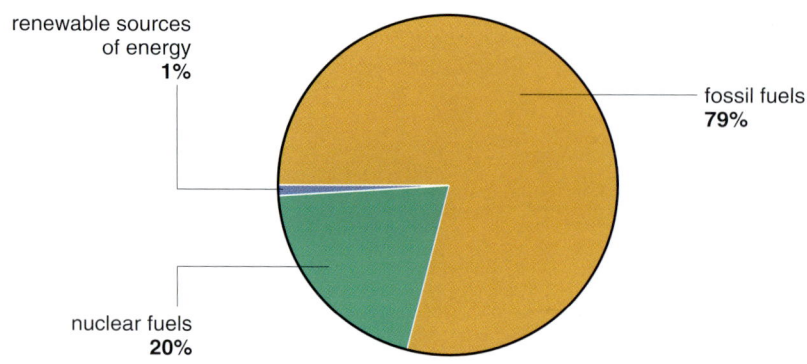

renewable sources of energy
1%

fossil fuels
79%

nuclear fuels
20%

Figure 9 ▲

How are fossil fuels formed?

Fossil fuels are formed from dead plants and animals which lived on the Earth millions of years ago. Those that died and sank to the bottom of lakes and seas became covered with layers of mud. Over a very long period of time, more and more layers formed, creating enormous pressures which gradually changed the dead plants and animals into coal, oil and gas.

Millions of years ago, plants and animals in the sea died and fell to the sea bed

They were covered by layers of mud and rocks. Over a long period of time, more and more of these layers built up

Enormous pressures were created as the layers sank deeper and deeper. These pressures changed the dead plants and animals into coal, oil or gas

Figure 10 ▲   The formation of fossil fuels

A **fuel** is a substance which releases energy when it is burned. The most common fuels include coal, oil, gas and wood. We use large quantities of fuel to generate electricity at our power stations.

To avoid running out of fossil fuels, we must use them more efficiently and we must look for alternative sources of energy that are **renewable**.

Test Yourself

12 What is a fuel?

13 Why are coal, oil and gas called non-renewable fuels?

14 Name one fuel which is renewable.

Renewable sources of energy

Wind energy

Heat energy from the Sun sets up convection currents in the Earth's atmosphere. The kinetic energy of the moving air can be used to turn the blades of wind turbines and so produce electricity. This can be a very useful source of energy, particularly for isolated communities with no national electrical supply. The energy supply is, however, intermittent – if there is no wind, there is no electricity.

Hydroelectric energy

Rainwater is stored behind dams. When it is released it drives turbines that generate electricity. This is a clean source of energy that creates little or no pollution. There are, however, two large disadvantages: a) the very high initial costs of building the dams and the generator plant and b) the flooding of large areas of land, destroying wildlife and their habitats.

Figure 11 ▲ Wind farms make use of the kinetic energy from moving air

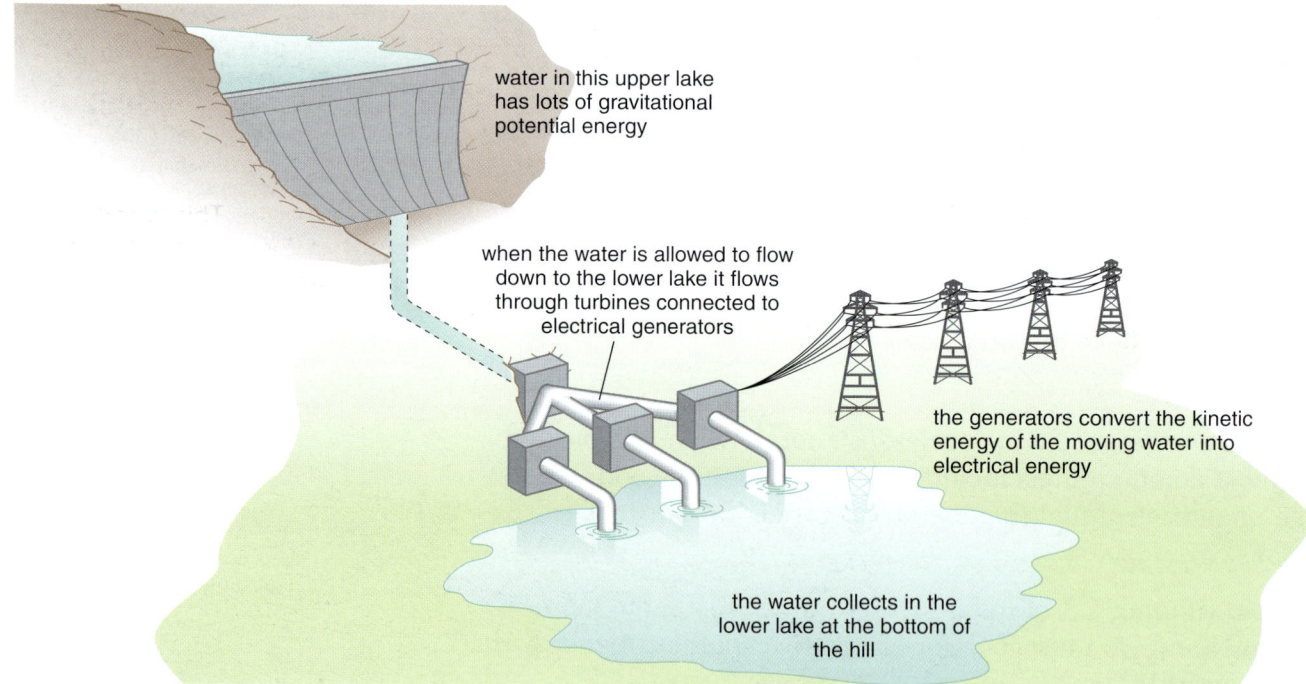

water in this upper lake has lots of gravitational potential energy

when the water is allowed to flow down to the lower lake it flows through turbines connected to electrical generators

the generators convert the kinetic energy of the moving water into electrical energy

the water collects in the lower lake at the bottom of the hill

Figure 12 ▲ Hydroelectric power plants make use of the gravitational potential energy of stored water

Tidal energy

At high tide, the tidal barrier is closed. As the water on the side nearest the sea falls, the water behind the barrier is held back. At low tide this water is released and used to drive turbines and produce electricity.

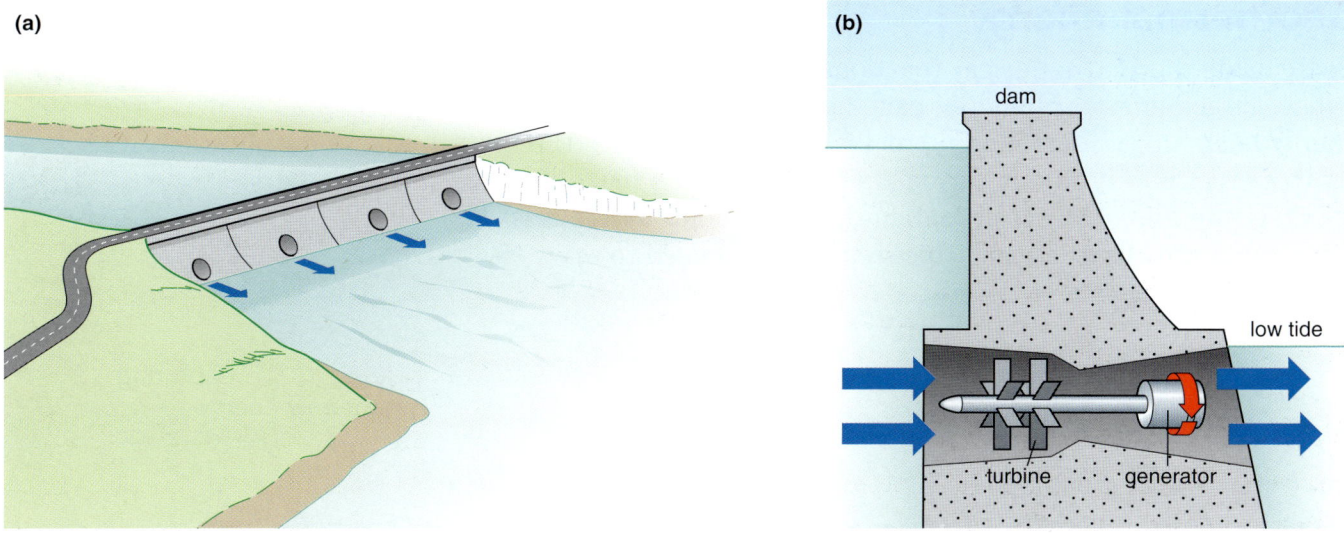

(a) (b)

Figure 13 ▲ Tidal barriers also use the gravitational potential energy of stored water

Wave energy

The constant movement of the waves across the surface of seas and oceans can be captured using devices similar to the 'duck' shown in Figure 14. As a water wave moves under the duck, it is made to move up and down. This simple motion can be used to generate electricity. This source of energy produces no pollution but very large areas of these machines are needed to collect enough energy to justify the cost of their construction.

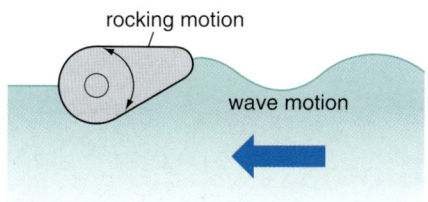

Figure 14 ▲ This 'duck' captures the movement energy from waves

Solar energy

Most of the Earth's energy comes from the Sun. **Solar cells** convert some of this energy directly into electricity and can be used to power small devices like calculators.

Some buildings, like the house in Figure 15, use **solar panels** to capture the Sun's energy and reduce their heating costs. Black surfaces within the panels absorb radiant energy from the Sun (infra red radiation). This in turn is absorbed by water flowing through pipes attached to these surfaces.

Figure 15 ▲

Geothermal energy

Deep inside the Earth, nuclear reactions are taking place which release large amounts of energy, called **geothermal energy**. This energy heats up the surrounding rocks. In certain regions, the Earth's crust is very thin or has faults (cracks) which make it possible to pump cold water deep under the ground where it is heated and returns as steam. This steam is then used to drive turbines and generate electricity.

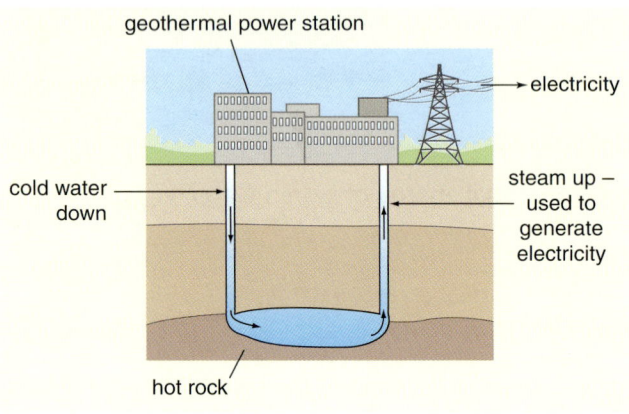

geothermal power station

electricity

cold water down

steam up – used to generate electricity

hot rock

Figure 16 ◄ Capturing geothermal energy

Biomass

Biomass measures the amount of any material that is or has been alive. As trees and plants grow, they increase their biomass.

Simple substances found in the soil and the atmosphere combine chemically by a process called **photosynthesis**.

> **water + carbon dioxide** $\xrightarrow{\text{sunlight}}$ **sugar (food for plant) + oxygen**

The oxygen released by this reaction escapes into the atmosphere. Some of the sugar or glucose may be used by a plant straight away to produce the energy it needs to live and grow. Often the glucose is used to make more complicated substances that increase the biomass of the plant, such as cellulose, which is used to make the walls of plant cells, and starch which is stored by the plant for later use.

There are many different ways in which these concentrated sources can be treated in order to release their energy.

1 These sources may be eaten:

Leaf → earthworm → shrew → owl

The earthworm obtains the energy it needs to live by eating leaves. The shrew obtains its energy needs by eating earthworms and the owl obtains its energy by eating shrews. In each case, energy is passed along the food chain.

2   These sources may be burned – we burn coal and wood to heat our homes.

3   These sources may be converted into other forms before they are used e.g. by fermentation. For example, some cars can run on a fuel created by fermenting the sugar from sugar cane to produce ethanol, a kind of alcohol.

Energy from the Sun

Figure 17 ▼   Energy from the Sun is captured and used in many different ways

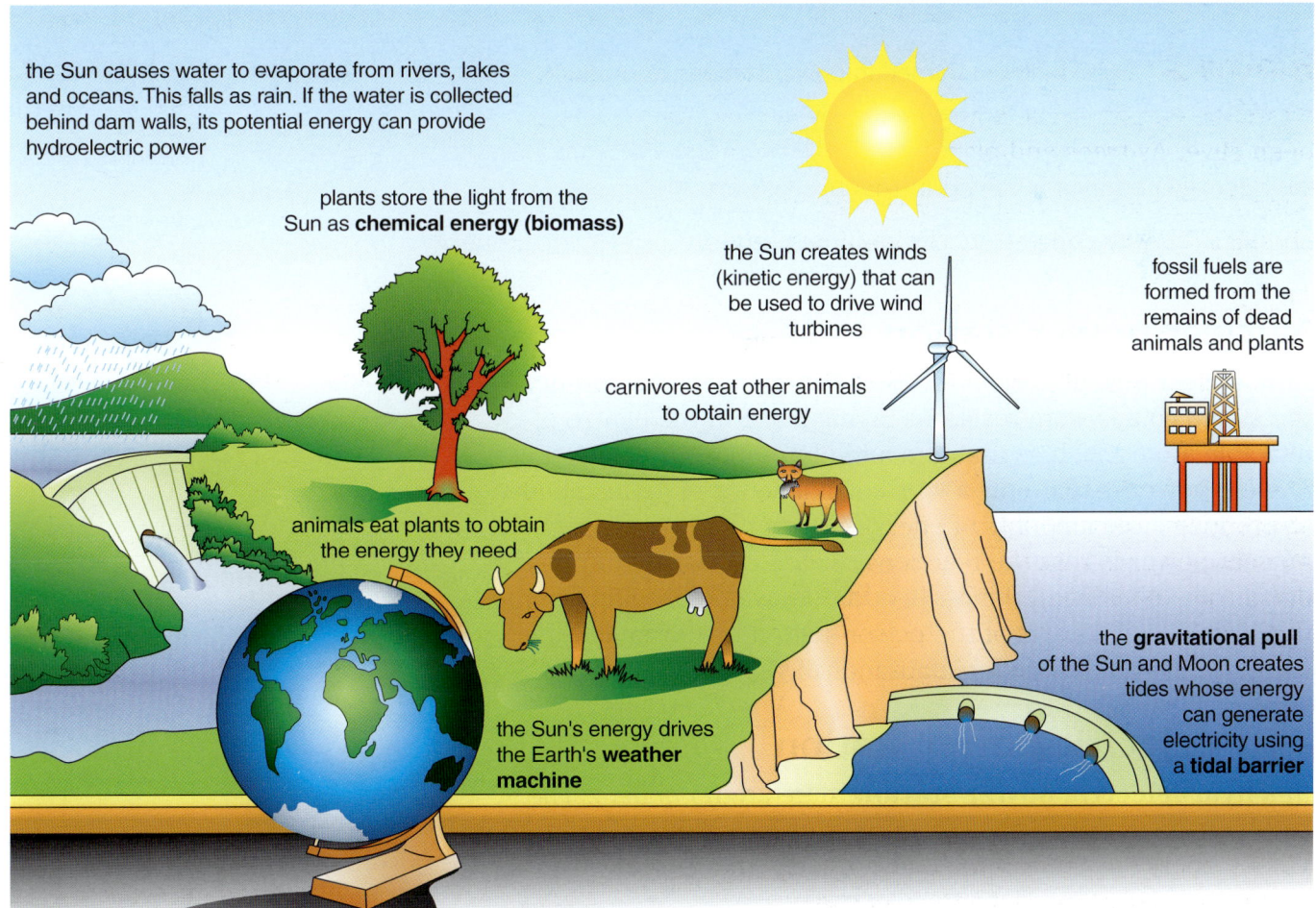

the Sun causes water to evaporate from rivers, lakes and oceans. This falls as rain. If the water is collected behind dam walls, its potential energy can provide hydroelectric power

plants store the light from the Sun as **chemical energy (biomass)**

the Sun creates winds (kinetic energy) that can be used to drive wind turbines

fossil fuels are formed from the remains of dead animals and plants

carnivores eat other animals to obtain energy

animals eat plants to obtain the energy they need

the Sun's energy drives the Earth's **weather machine**

the **gravitational pull** of the Sun and Moon creates tides whose energy can generate electricity using a **tidal barrier**

Generating electricity

Electrical energy is one of the most convenient forms of energy as it is easy to transform it into other types of energy. Most of the electricity we use at home is generated at a power station

Fossil fuel power station

At a coal, oil or gas power station, fuel is burned to release heat energy. This energy is then used to heat water, changing it into steam. The energy contained by the steam is then used to turn turbines. The turbines turn **generators** which produce electricity. The electrical energy produced travels along a network of wires called the National Grid to our homes.

Test Yourself

18 Explain how energy from the Sun has been given to the water stored behind dams.

19 Explain how the energy stored in fossil fuels originally came from the Sun.

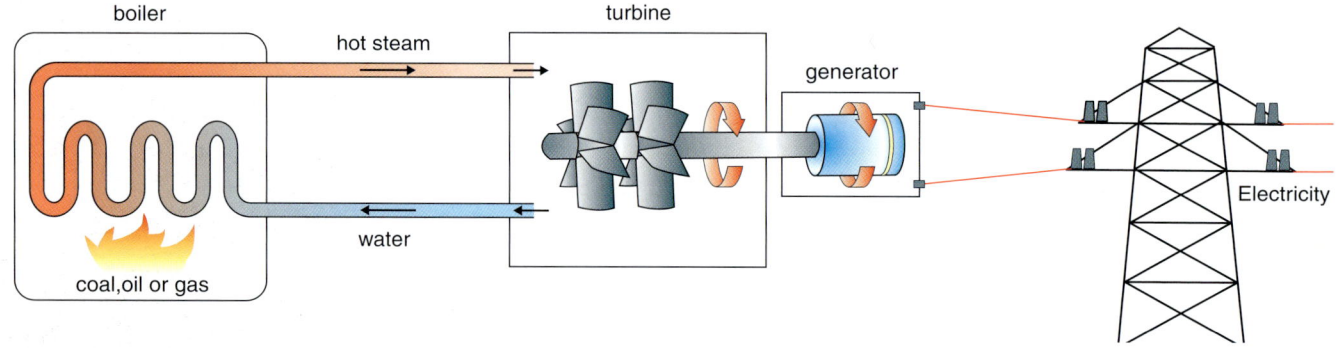

Figure 18 ▲   Fossil fuels are burned in power stations to produce electricity

The energy changes involved in this process are:

chemical energy → heat energy → kinetic energy → electrical energy

Nuclear power station

At a nuclear power station, the heat energy needed to produce the steam comes from reactions taking place in the centre of unstable atoms such as uranium and plutonium. One advantage of nuclear power is that only a very small amount of fuel is needed to produce large amounts of energy. The big disadvantage of nuclear power is that the reactions release radiation which is dangerous to all living things. In addition, waste products from the reactions and parts of the power station itself remain dangerously radioactive for thousands of years.

Hydroelectric power station

In a **hydroelectric power station** there are no boilers. The kinetic energy of the falling water is transferred directly into the kinetic energy of the turbine blades. The turbines then turn the generators, which produce the electricity.

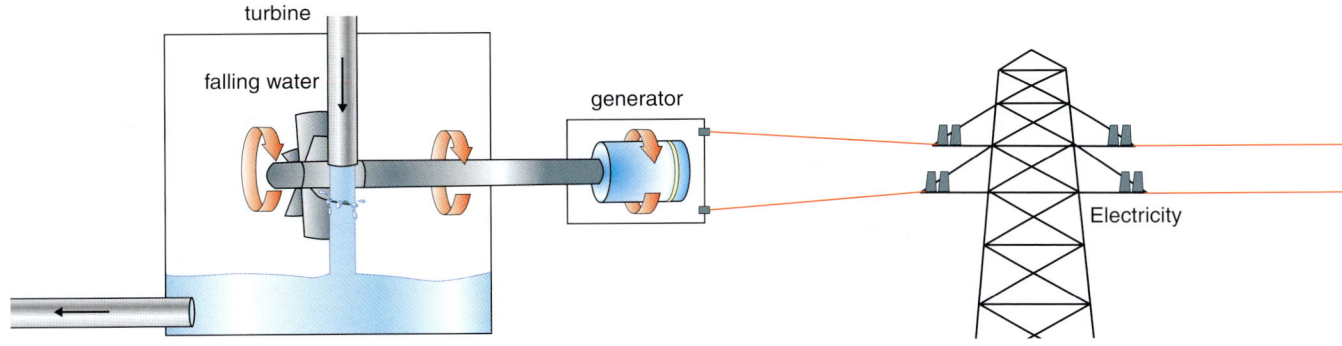

Figure 19 ▲ The workings of a hydroelectric power station

The energy changes involved are:

gravitational potential energy → kinetic energy of water → kinetic energy of turbines → electrical energy

Figure 20 ▲ An electricity meter

Test Yourself

20 Why is electrical energy a convenient form of energy?

21 How is the energy stored in fossil fuels released at a power station?

22 What drives the turbines at a hydroelectric power station?

23 What do the turbines turn at a power station in order to produce electricity?

Paying for your electricity

As the electrical energy from a power station enters our homes, it passes through an 'electricity meter' which measures how much electrical energy we have used. The meter measures this energy in **units** or **kilowatt-hours**. Every 3 months we receive a bill for the electrical energy we have used.

Most appliances in our homes need electrical energy to work, but they use the energy at different rates. Some, like electric heaters, use the energy quickly. Others, like light bulbs, use the electrical energy much more slowly. The rate at which energy is being used is called the **power** or **power rating** of the appliance. We measure power in **watts (W)** or **kilowatts (kW)**. One kW is 1000 W.

Light bulbs can be bought with different wattages depending on where they are to be used. A 60 W bulb is brighter than a 10 W bulb because it is converting the electrical energy into heat and light energy six times faster.

Figure 21 ▲ The power rating of most domestic electrical appliances is shown on a label attached to their casing

Appliance	Typical power rating	Appliance	Typical power rating
TV set	600 W	electric iron	500 W
kettle	2 kW	washing machine	2 kW
large electric heater	3 kW	stereo system	80 W
hair drier	1200 W	bedroom light	60 W

Table 3 ▲ Power ratings for some domestic appliances

Test Yourself

24 In what units do we measure the power of an appliance?

25 How can we find out how much electrical energy we will use in the next three months?

26 Suggest an approximate power rating for the following:
a) a dim light bulb
b) a very bright light bulb
c) a hair drier
d) a large heater.

Summary

When you have finished studying this chapter, you should understand that:

✔ There are many different kinds of energy including light energy, heat energy, sound energy, electrical energy, chemical energy, nuclear energy, strain and gravitational potential energy, kinetic energy and nuclear energy.

✔ Energy allows us to live and work.

✔ When energy is used, it may change into other forms. Some of these forms may be dilute and the energy difficult to reuse.

✔ If a device changes a large amount of its input energy into the required form of energy, the device is efficient.

✔ Coal, oil and gas are fossil fuels. They are non-renewable sources of energy. In order to conserve fossil fuels we must make more use of renewable sources of energy such as wind, hydroelectric, tidal, wave, solar, biomass and geothermal.

✔ Practically all the energy we have on the Earth comes from or came from the Sun.

✔ Electrical energy is the most convenient form of energy. Most of the electrical energy we use is produced at power stations by generators.

✔ Different appliances have different power ratings i.e. they transform energy at different rates. We measure power rating in watts.

End-of-Chapter Questions

1 Explain in your own words the following key terms you have met in this chapter:

kinetic energy

elastic or strain potential energy

gravitational potential energy

nuclear reaction

energy transformation

dissipated

concentrated and dilute forms of energy

efficiency

fossil fuels

non-renewable source of energy

fuel

renewable source of energy

solar cell

solar panel

geothermal energy

biomass

photosynthesis

generator

hydroelectric power station

units/kilowatt-hour

power/power rating

watt/kilowatt

2 What device would you use to change

a) sound energy into electrical energy?

b) electrical energy into sound energy?

c) electrical energy into heat and light energy?

d) light energy into electrical energy?

e) strain potential energy into kinetic energy?

f) electrical energy into gravitational potential energy?

3 a) What is a fuel?

b) Name three fossil fuels.

c) Name one renewable fuel.

d) Name one renewable source of energy that

   i)   can be used in any weather

   ii) has very high initial costs

   iii) might cause noise pollution

   iv) has a large effect on the environment.

4 Over the next 24 hours make a list of everything you eat (and drink). Then calculate the total energy this food has provided for you. Compare this value with the bar chart on page 4.

5 Explain how energy from the Sun is transformed into

a) wind energy

b) fossil fuels

c) hydroelectric energy.

6 The diagram on page 16 shows a pumped storage power station. During the daytime, water is released from the upper lake and is used to drive the turbines to generate electricity. During the night-time, electrical energy from another power station is used to pump some of the water back up to the upper lake.

a) What kind of energy does the water in the upper lake have?

b) What kind of energy does the water have after it has been released and is about to pass over the turbines?

c) What happens to the generator as water passes over the turbine blades?

d) Why is water only released from the top lake during the daytime?

e) What kind of power station might be used to provide the electrical energy to pump the water uphill during the night? Explain your answer.

f) Give one advantage and one disadvantage of this type of energy source.

End-of-Chapter Questions continued

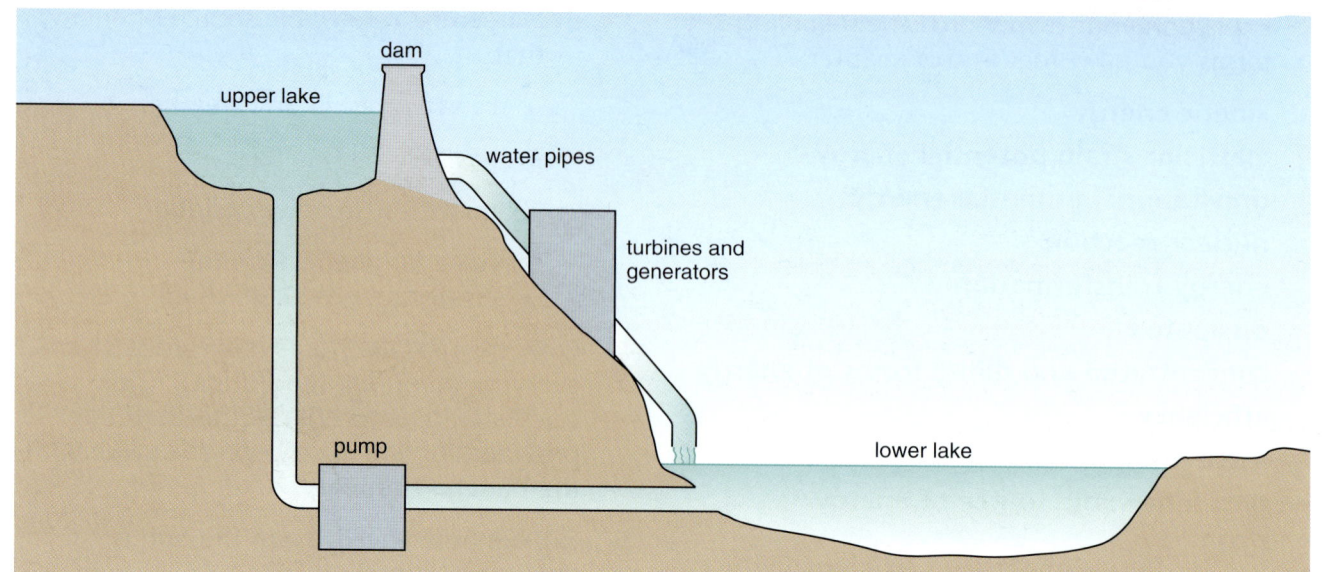

7   Try to find an old electricity bill for your house. How many units of electricity did you use in 3 months? Would you expect to use more or less energy in the summer months or in the winter months? Explain your answer.

8   Try to find out the power rating of 10 electrical appliances in your home. Put the 10 into a list beginning with the most powerful. Look carefully at your list. Can you see a pattern?

2 Forces and their effects

This Olympic weightlifter is going to have to apply a very large force to these weights in order to lift them over his head. When you have finished studying this chapter you should be able to say exactly how large this force must be . . . or perhaps you already know?

Different types of force

There are many different types of force that can be applied to an object. These include pushes, pulls, twists, turns, stretching forces and squashing forces.

Test Yourself

1   Write five sentences which describe five different types of force you have applied to objects today.

Effects of forces

If we apply a force to an object, it is likely that it will do one of the following.

- Cause an object to speed up or **accelerate**.
- Cause an object to slow down or **decelerate**.
- Cause a moving object to stop.
- Cause a stationary object to start moving.
- Change the direction in which an object is moving.
- Change the shape of an object.

Figure 1 ▶   Can you work out which picture relates to which type of force?

Forces without contact

Sometimes it is not necessary to be in contact with an object in order to apply a force to it.

Look carefully at the two pictures below.

Test Yourself

2 Describe six examples where each of the effects of a force can be seen in sport.

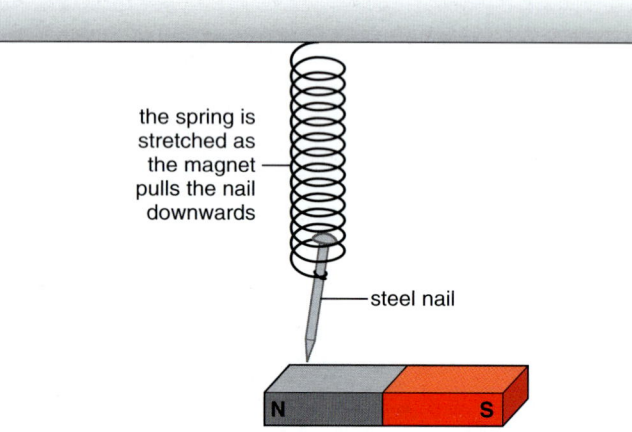

the spring is stretched as the magnet pulls the nail downwards

steel nail

N   S

Figure 2 ▲ a) This steel nail is being pulled downwards by the **magnetic attraction** of the bar magnet

There are gravitational forces of attraction between all objects but they are often very weak. They are only really noticeable when one or both of the objects is massive, such as a planet, moon or star. It is gravitational forces of attraction that hold the planets in orbit around the Sun (see page 124).

Figure 2 ▲ b) This bungee jumper is being pulled downwards by the **gravitational attraction** between her and the Earth. This attractive force between an object and the Earth we call **weight**

Mass and weight

Although a space traveller weighs less on the Moon than he does on Earth, he still contains the same amount of matter. His **mass** is the same. Mass is the amount of matter in an object and is measured in kilograms (kg). Weight is a force caused by gravity acting on a mass and is measured in newtons (N).

Object	Typical mass	Weight on Earth	Weight on Moon
bag of sugar	1 kg	10 N	1.6 N
apple	0.1 kg	1 N	0.2 N
sack of potatoes	25 kg	250 N	42 N
small boy	40 kg	400 N	67 N
large man	100 kg	1000 N	160 N

Table 1 ▲ Examples of masses and weights

Figure 3 ▲ 'There is an attractive force between me and the Earth. This force is my weight. The Moon is much smaller (less mass) than the Earth so the pull of its gravity on me is less (approximately 1/6th), therefore on the Moon I weigh much less.

Oh no! Jupiter is much larger (more mass) than the Earth so the pull of gravity is larger (approximately 3 times larger). On Jupiter I weigh a lot, lot more. Don't come here to lose weight!'

Test Yourself

3 Name two forces that can be applied to an object without being in contact with it.

4 Explain the difference between the mass of an object and its weight.

5 Why would an object on Saturn weigh more than the same object on the Earth?

Read the following account carefully. Then answer Questions 6, 7 and 8

Colonel Black woke up and unzipped his sleeping bag. Almost immediately, he started to drift away from his sleeping station. Before he had gone too far, he used the rear wall to propel himself to the far side of the cabin where he stored his personal effects. Using his hands to bring himself to rest he opened a cupboard and removed his toothbrush and paste. He removed the screw top from the tube and eased some of the paste onto the bristles. Very carefully, so as not to lose any of the paste, he started to clean his teeth. The morning wash that followed was less difficult but not as satisfying as the shower he would have taken at home. The body wash with the large damp cloth and towel provided was certainly as hygienic but definitely not as refreshing. Having lived in the orbiting space station for 2 weeks, Black was beginning to get accustomed to his daily routine but he didn't think he would ever get used to the effects of weightlessness.

6 Identify five occasions when Colonel Black applied a force. Describe the effect of each of these forces.

7 Explain why it would be difficult for Colonel Black to take a shower whilst on the space station.

8 What do you understand by the word 'weightlessness'? Where could someone experience the effects of weightlessness? Write down two advantages and two disadvantages of being weightless.

Measuring forces

We measure the size of a force in units called **newtons** (N). If you hold a medium-sized apple in your hand, it will be pushing down with a force of approximately 1 N.

The sizes of some other forces are shown in Figure 4.

Figure 4 ▲   Comparing the sizes of different forces

Test Yourself

9   Estimate the sizes of the forces you might apply in the following situations:
   a) picking up a newspaper
   b) pulling a string to turn on the light in a bathroom or bedroom
   c) pushing an average sized family car along a flat road
   d) snapping an elastic band
   e) opening a car door
   f) lifting a cup of coffee.

The units of force are named after one of history's most famous scientists, Sir Isaac Newton.

Sir Isaac Newton

Figure 5 ▲ Sir Isaac Newton

- The development of his three Laws of Motion on which the study of mechanics is based.
- His study of circular motion and its application to the movements of the Moon. These investigations eventually led to his theory of gravity and the Universal Law of Gravitation. It is this law which explains and predicts the movements of all heavenly bodies.
- The invention of calculus (an extremely useful mathematical process) as a consequence of which, he was recognised as being one of the most noted mathematicians in Europe at that time.

In 1703, he was made President of the Royal Society and in 1705, he was the first British scientist to receive a knighthood for his research. He died in March 1727 at the age of 84.

Sir Isaac Newton was born in Lincolnshire on 4 January 1643. He went to Trinity College, Cambridge to study law. Around this time a scientific revolution was taking place led by great scientists such as Copernicus, Kepler and Galileo. Fascinated by these developments and some of the explanations that were being given, Newton developed some theories of his own that were to shake the world of science. Among his most noted contributions were:

- His discovery that white light is a mixture of coloured lights (see Chapter 9).
- The invention of the reflecting telescope.

Find out

1 Who were Copernicus, Kepler and Galileo? What contributions did they make to science in the 17th Century?

2 Why did Newton believe that white light was actually lots of coloured lights mixed together?

3 What is a refracting telescope? What is a reflecting telescope? What is the main advantage of using a reflecting telescope compared with a refracting telescope?

4 Write down Newton's Three Laws of Motion and his Universal Law of Gravitation.

5 What is the Royal Society?

Measuring forces

We can measure the size of a force using a **newtonmeter**.

A newtonmeter contains a spring. When a force is applied to the spring it stretches. The larger the force, the more the spring stretches. A scale and pointer show the size of the force.

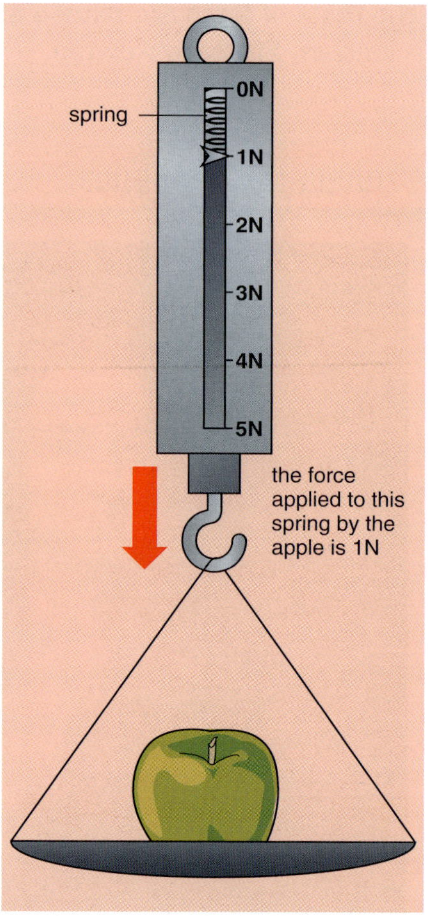

Test Yourself

10 The diagram below shows five different newtonmeters.

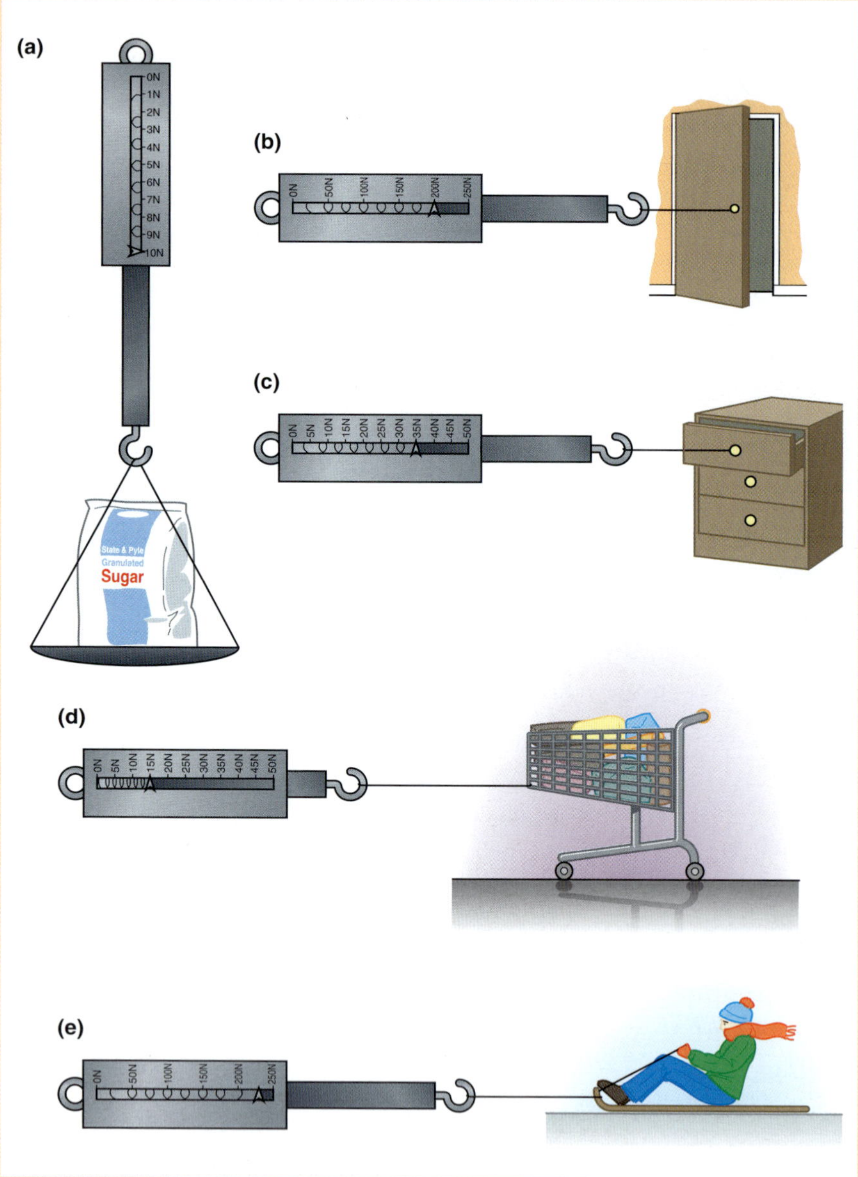

(a)

(b)

(c)

(d)

(e)

a) Read the size of the force which is being applied to each of the newtonmeters.
b) Suggest one difference between the springs in the first two newtonmeters.

Figure 6 ▲ The reading on this newtonmeter is 1 N

spring

0N
1N
2N
3N
4N
5N

the force applied to this spring by the apple is 1N

Making a simple newtonmeter

A spring with a pointer attached is suspended next to a rule whose scale has been covered with plain tape. The position of the pointer when no force is applied to the spring is marked on the tape. A selection of forces are then applied to the spring. The position of the pointer and the value of each of the applied forces are recorded on the tape. These marks and their values form a rough scale on the tape. If a force whose value is not known is applied to the spring, its size can be read from the scale.

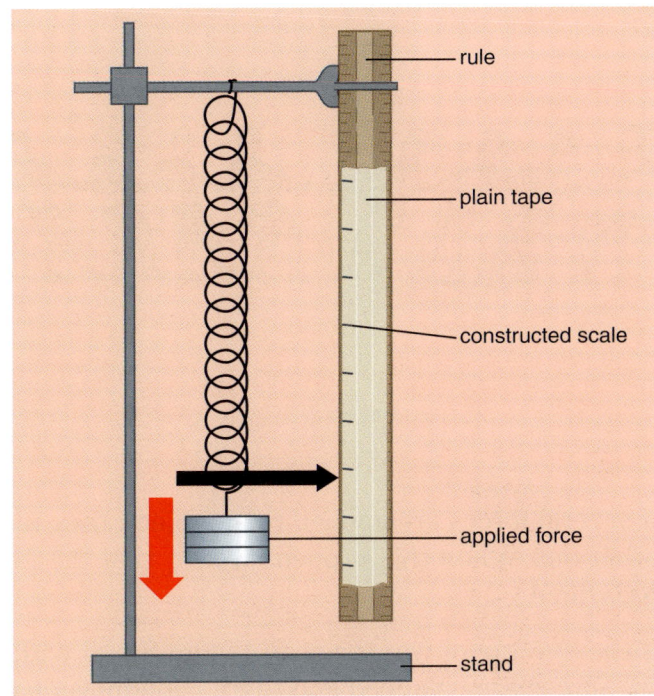

- rule
- plain tape
- constructed scale
- applied force
- stand

Figure 7 ▶ How to make a simple newtonmeter

Extension box

A pupil carried out an experiment to calibrate a newtonmeter. He hung weights on a spring and then measured the **extension** they created. The results of the experiment are shown in the table below.

Extension (mm)	1.5	3.0	6.0	7.5	9.0	12.0
Force (N)	2.0	4.0	6.0	10.0	12.0	16.0

Using the above results, draw a graph of the extension of the spring (y-axis) against force applied to spring (x-axis).

From your graph, determine a) the forces, which will cause the spring to extend by 4.0 mm and 10 mm, and b) the extensions of the spring when a force of 5.0 N and 8.5 N are applied to it.

The graph you have drawn is a straight-line graph that passes through the origin (0,0). The shape of the graph indicates the relationship between the two quantities that have been plotted. A straight-line graph which passes through the origin indicates that there is a **direct proportionality** between the two quantities which have been plotted i.e. if one of the quantities is doubled, so too is the other. In this case, it shows that if the force applied to the spring is doubled, the extension is doubled. If the force applied to the spring is trebled, the extension is trebled etc.

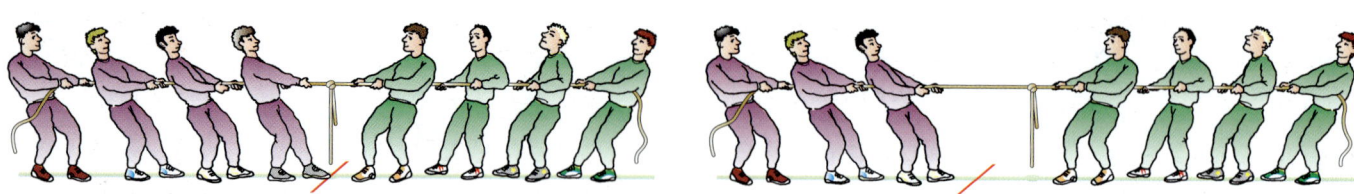

Figure 8 ▲   Figure 9 ▲

Balanced and unbalanced forces

If several forces are applied to an object, they may cancel each other out. When this happens we say that the forces are **balanced**.

If the two tug of war teams in Figure 8 pull with the same force, the forces are balanced and there is no movement.

If one of the teams pulls with a force which is greater than that of the opposition, as in Figure 9, the forces are **unbalanced** and there is movement.

The ship in Figure 10 is stationary. There are several forces acting upon it, so these forces must be balanced. Gravitational forces, i.e. the weight of this ship, are pulling it down. But a second force called the **upthrust** of the water is pushing it up. These two forces are equal and balanced so the ship floats. If the weight of the ship was increased by adding more cargo, the ship would sink deeper into the water. This would create more upthrust, which would balance the extra weight.

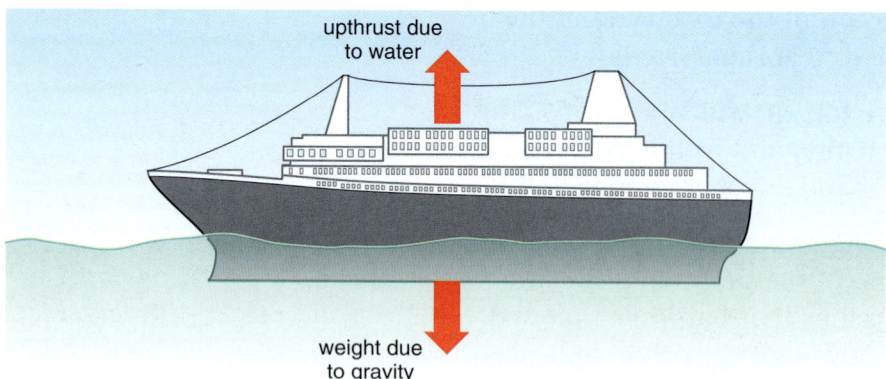

Figure 10 ▲   The upthrust of the water helps this ship to float

Most ships have lines drawn on their sides called **plimsoll lines**. These indicate when a ship is carrying the maximum safe weight of cargo. If more cargo is added, the ship will sit too low in the water.

Test Yourself

11   Which of the following are examples of balanced forces:
  a) a box on a table
  b) a floating ship
  c) a braking car
  d) an aeroplane taking off
  e) a stationary weight hanging from a spring?

12   What is upthrust?

Friction

One of the most common forces which acts upon an object is **friction**. Friction is present whenever an object moves or tries to move. It acts in the direction which opposes the motion.

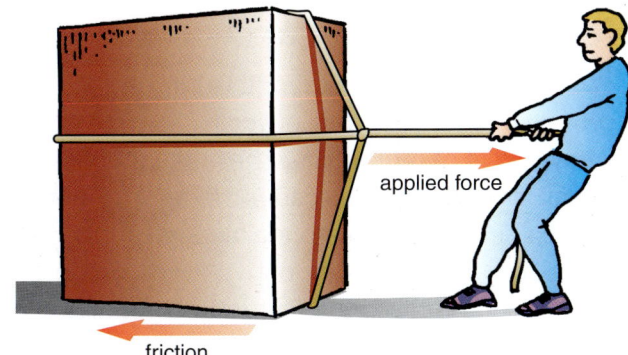

applied force

friction

Friction always acts in the direction which opposes the motion.

Figure 11 ▲ The friction between the crate and the ground makes it difficult to pull

On some occasions friction can prove very useful. For example when you walk or run, you push yourself forward by pushing backwards on the ground. Friction between your foot and the floor helps you to do this. If there was no friction, your feet would slip.

Smooth surfaces reduce the friction between objects while rough surfaces increase frictional forces. Trainers and football boots are designed to prevent your foot from slipping by increasing the frictional forces between you and the ground. In contrast, ice skates and skis are designed with smooth surfaces which keep friction to a minimum.

Whenever an object moves through the air it experiences frictional forces or **drag** which try to prevent its motion. The faster the object moves, the greater the drag. To reduce these forces, objects like a bobsleigh are shaped so that they cut through the air. They are **streamlined**. To decrease the friction between the ice and the bobsleigh, the runners are coated with a **lubricant** such as wax.

Figure 12 ▲ This bobsleigh is streamlined to reduce drag

Test Yourself

13 What happens to the frictional forces experienced by an aircraft if it increases its speed?

14 Suggest one way in which the designers of aircraft try to keep drag forces to a minimum.

15 What is a lubricant?

Friction and the car

Most modern cars are streamlined in order to reduce friction forces from the air. But in order to control a car's speed and direction, it is essential that there is lots of friction between the tyres of the car and the road surface. A new tyre will provide good grip as it has a rough surface with lots of tread. A worn tyre with less tread and a smoother surface will provide less grip and the car will be much more difficult to control and stop.

Figure 13 ◄ The new tyre has more tread than the worn tyre, so grips the road better

Having tyres which are in good condition is important if you need to stop quickly but there are several other important factors which will affect how quickly a car driver can stop. These are

- the reactions of the driver
- the braking system of the car
- the speed of the car
- the weather/road conditions.

The diagrams below show how the shortest possible stopping distances under ideal conditions increase dramatically with speed.

at 13 m/s (30 mph)

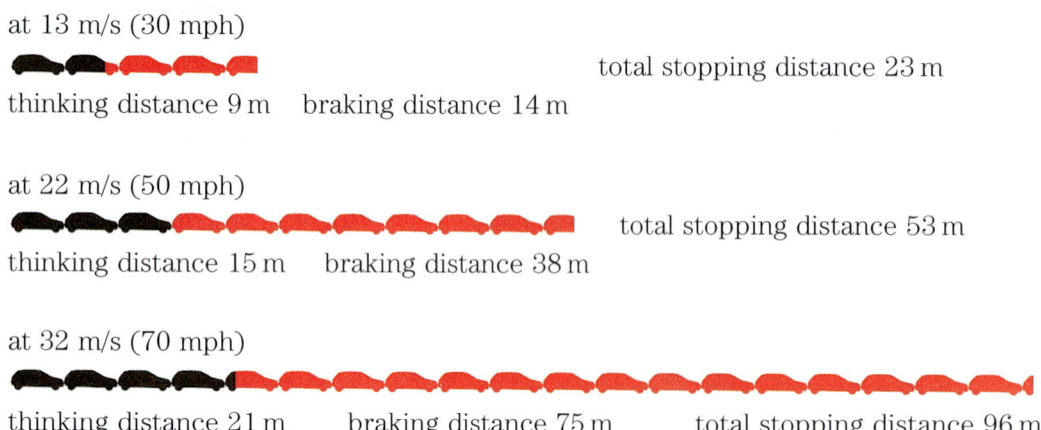

total stopping distance 23 m

thinking distance 9 m   braking distance 14 m

at 22 m/s (50 mph)

total stopping distance 53 m

thinking distance 15 m   braking distance 38 m

at 32 m/s (70 mph)

thinking distance 21 m   braking distance 75 m   total stopping distance 96 m

Summary

When you have finished studying this chapter, you should understand that:

✔ Pulls, pushes and twists are all different kinds of force.

✔ Forces may cause an object to change its motion or its shape.

✔ The size of a force is measured in newtons (N).

✔ We measure the size of a force using a newtonmeter.

✔ The extension of a spring is directly proportional to the force applied to it.

✔ Weight is the force exerted on an object due to gravity.

✔ Mass is the amount of matter in an object.

✔ Balanced forces produce no change to motion.

✔ Unbalanced forces cause changes to motion.

✔ Friction is a force that opposes motion.

End-of-Chapter Questions

1 Explain in your own words the following key terms you have met in this chapter:

accelerate
decelerate
magnetic attraction
gravitational attraction
weight
mass
newton
newtonmeter
extension
direct proportionality
balanced forces
unbalanced forces
upthrust
plimsoll line
friction
drag
streamlined
lubricant

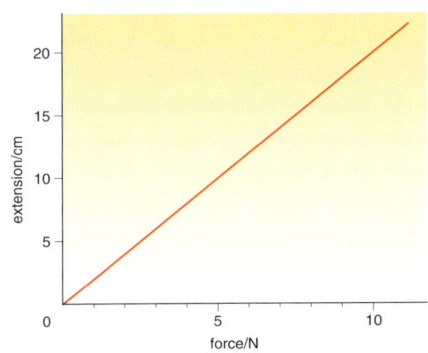

2 The graph on the left shows how a spring extends when forces are applied to it.

a) What force is necessary to cause the spring to extend by 5.0 cm?

b) By how much would the spring extend if a force of 10.0 N was applied to it?

The graph is a straight line passing through the origin.

c) What happens to the amount by which the spring stretches if the force applied to it is doubled?

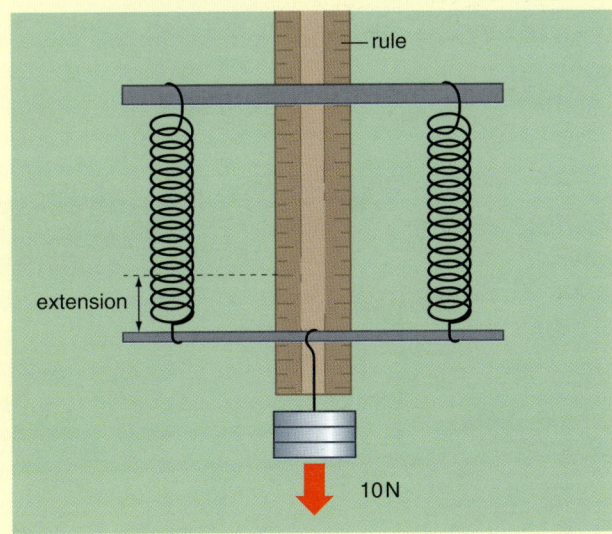

d) If a weight of 10 N is hung from both springs, as shown in the diagram, by how much would each spring stretch?

End-of-Chapter Questions continued

3 The gravitational field strength on the Earth is 10 N/kg. This means that an object which has a mass of 1 kg will weigh 10 N on the Earth.

a) What is the weight on Earth of an object which has a mass of 5 kg?

The gravitational field strength on Jupiter is 26 N/kg?

b) What is the weight on Jupiter of the 5 kg object?

Gravitational forces can be applied to an object without being in contact with it.

c) Name one other kind of force that can be applied to an object without being in contact with it.

4 The diagram below shows the four forces experienced by a ship as it sails.

a) Describe the sizes of these forces in the following situations:

i)   The ship is travelling on the surface of the water at a constant speed.

ii) The ship is travelling along the surface of the water and accelerating.

iii) The ship is stationary but starting to sink.

b) Give one word to describe the shape of the ship.

c) Why does the ship have this shape?

d) Draw a picture of a dolphin and explain why it has this shape.

5 Find out how it is possible to start a fire using just two pieces of wood and some dry straw.

6 The Highway Code tells us the stopping distance of cars travelling at different speeds.

a) Find out the stopping distance for a car travelling at 30 mph (48 kph) and 70 mph (108 kph).

b) Name two things that might affect the stopping distance of a car apart from its speed.

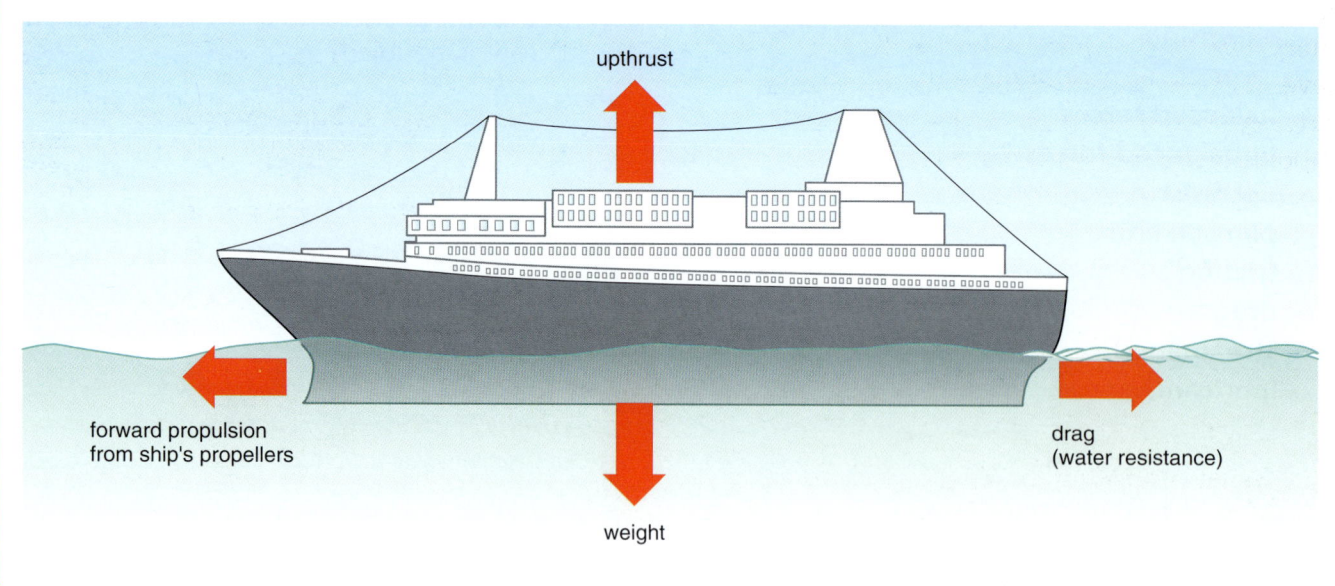

upthrust

forward propulsion
from ship's propellers

drag
(water resistance)

weight

Electrical circuits

If we build the circuit shown, an **electric current** will flow around the circuit and the bulb will glow. But what is an electric current and why does the bulb glow?

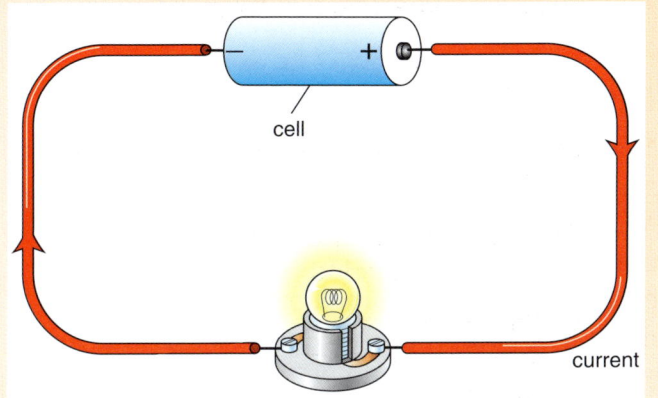

Simple circuits

An electric current is a flow of charge. The charges are made to move by the **cell**. We can imagine the cell behaving like a pump giving energy to the charges. If we want a larger current to flow and the charges to have more energy we can connect several cells together. Several cells connected together is called a **battery**. When cells are connected together it is important that we make sure they are all *pushing* in the same direction.

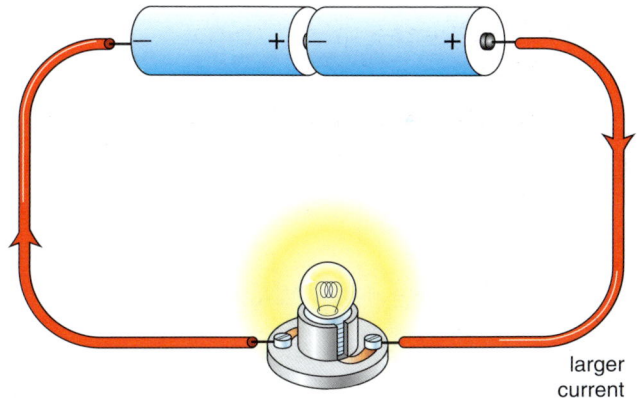

Figure 1 ▲   A simple two cell circuit

Test Yourself

1   What is an electric current?

2   What do cells do when they are connected in circuits?

3   What is a battery?

Like water flowing through pipes, electrical charges need something to travel through. In most circuits, like the one above, the 'pipes' for charges are metal wires. Metals are good **conductors** of electricity. They allow charges to flow through them easily. Materials such as plastics which do not allow charges to flow through them are called **insulators**.

If charges can flow all the way around a circuit and back to the cell, we say that the circuit is **complete**.

If any of the wires are removed or there is a gap in the circuit, it is described as being **incomplete**. Current will not flow in an incomplete circuit regardless of the position of the gap or break.

Circuit diagrams

Drawing circuits like those above is not easy. It is much simpler to draw **circuit diagrams**, which use symbols to represent the various **components** in a circuit. Components are the bits and pieces that are connected together to make a circuit. Table 1 shows the symbols for some of the most common components.

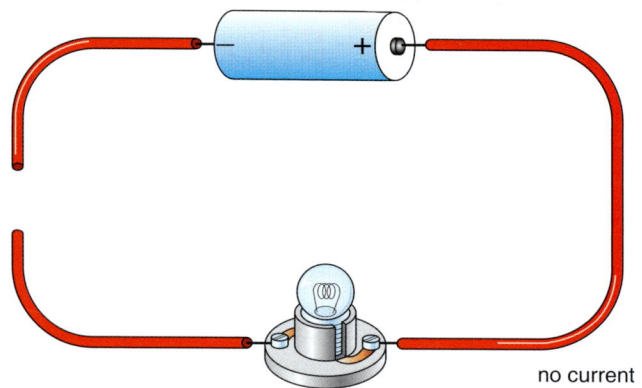

Figure 2 ▲   An incomplete circuit. No current flows and the bulb does not glow

What it is	What it looks like	Symbol	What it does
Cell			Pulls and pushes charges around a circuit.
Battery			Provides a larger current than a single cell.
Connecting wire			Provides a path through which current can flow.
Lamp/bulb			Glows brightly if sufficient current flows through it.
Switch			Turns current in a circuit on or off.
Resistor			Reduces the current flowing in a circuit.
Variable resistor			By altering the value of a variable resistor the size of the current can be changed.

Table 1 ▲   Symbols for some common components

Series and parallel circuits

There are two main kinds of electrical circuit. These are called **series circuits** and **parallel circuits**.

Figure 3 shows some examples of series circuits. Series circuits have no branches or junctions. The charges flow around one path.

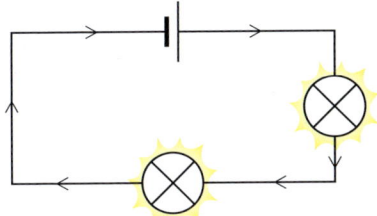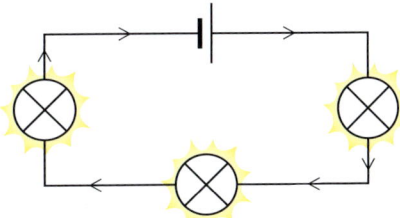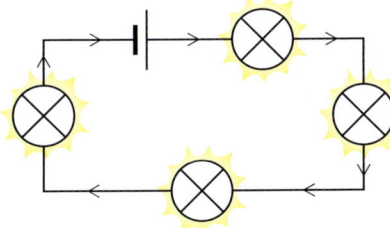

Figure 3 ▲   Series circuits

Figure 4 shows examples of parallel circuits. These circuits all have junctions and branches. The charges have more than one path along which they can flow.

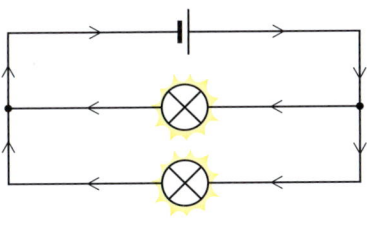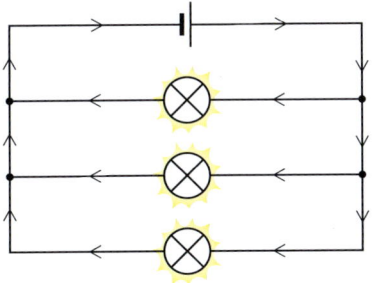

Figure 4 ◄   Parallel circuits

Switches

We use switches to turn circuits on and off by making them complete or incomplete. Switches in series circuits turn the whole circuit on and off. It is not possible to turn just one part of a series circuit on or off.

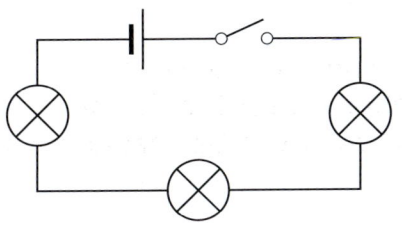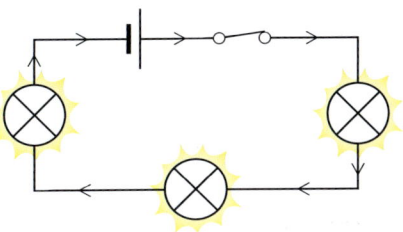

Figure 5 ◄   The whole of this series circuit is turned on or off using the switch

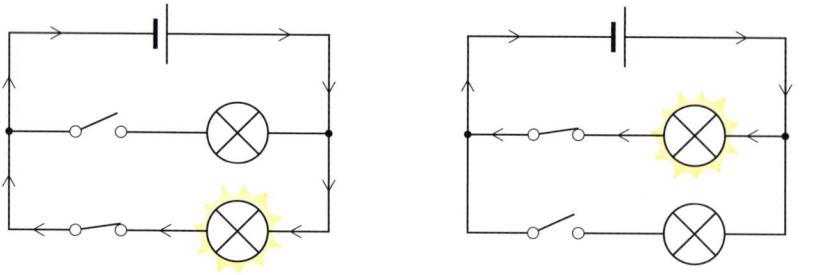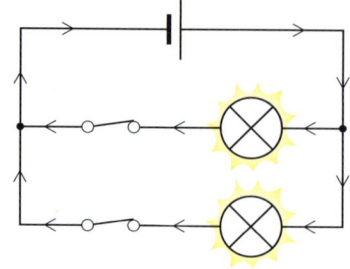

Figure 6 ▲   In a parallel circuit, switches can be used to turn parts of the circuit off

Switches in parallel circuits can be used to turn the whole circuit on and off or just part of it.

Figure 7 below shows a special circuit which contains 2 two-way switches.

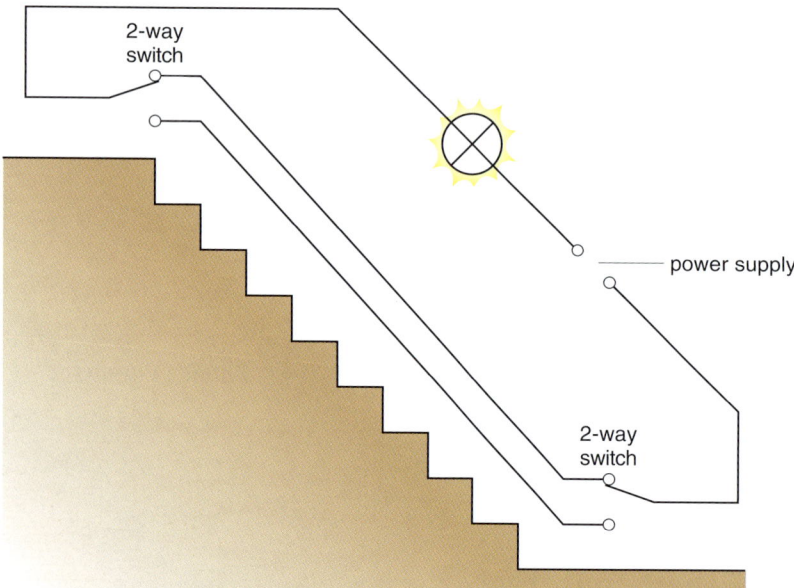

Figure 7 ▲   Using two-way switches in a lighting circuit

Using a circuit like this we are able to turn stair lights on and off from both the top and the bottom of the stairs.

Test Yourself

7   Explain in your own words the difference between a series circuit and a parallel circuit. Draw one example of each.

8   Explain why a circuit is turned off when a switch is opened.

Extension box

There are many other circuits which, like the circuit for the stair lights, contain two or more switches.

AND circuits

The circuit shown is called an AND circuit because for the circuit to work, switch A AND switch B have to be closed. This kind of circuit might be used to control the motors of a lift. One of the switches is closed when the outer door of the lift is shut. The second switch is closed when the inner door of the lift is shut. The circuit is now complete and the motor can operate the lift. If either or both of the doors is not shut, the motor will not operate. In this situation the AND circuit is being used as a safety device.

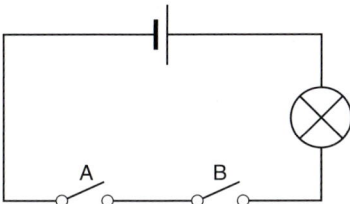

Figure 8 ▲ A circuit diagram for an AND circuit

We can represent all possible conditions of the switches and the circuit in a table called a truth table.

Switch A	Switch B	Circuit
open	open	off
open	closed	off
closed	open	off
closed	closed	on

Table 2 ▲ A truth table for an AND circuit

OR circuits

If two switches are connected in parallel they form an OR circuit.

It is given this name because for the circuit to work, switch A must be closed OR switch B must be closed OR both must be closed. This kind of circuit could be used to lower the barriers at a railway crossing and turn on warning lights and buzzers.

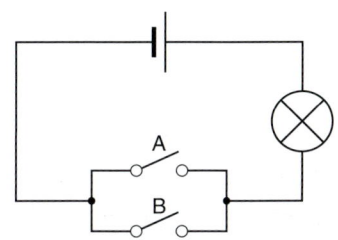

Figure 9 ▲ A circuit diagram for an OR circuit

Extension box continued

Switch A	Switch B	Circuit
open	open	off
closed	open	on
open	closed	on
closed	closed	on

Table 3 ▲ A truth table for an OR circuit

Extension box

Conventional current

When scientists first experimented with current electricity, they did not know which kinds of charges were moving around a circuit. They guessed that the charges were positive and therefore would flow from the positive side of a cell around a circuit and back to the negative side.

We now know that the charges that flow through wires are called electrons and that they carry a negative charge. To avoid too much change, scientists agreed they would continue to think of electric current as flowing from positive to negative. They called this **conventional current**.

Measuring current

We measure electric current in units called **amperes** (A) (often abbreviated to amps).

Electrical appliance	Typical current
household filament light	0.4 A
colour television	3 A
large electric fire	12 A

Table 4 ▲ Typical current for some household appliances

The size of a current is measured using an instrument called an **ammeter**. The ammeter is placed *in series* with the part of the circuit we are interested in.

We can see from Figure 10 that in a series circuit, the current leaving a cell or battery is the same as the current returning to it. We can see also that current is not 'used up' as it flows around a circuit. The size of the current in all parts of a series circuit is the same.

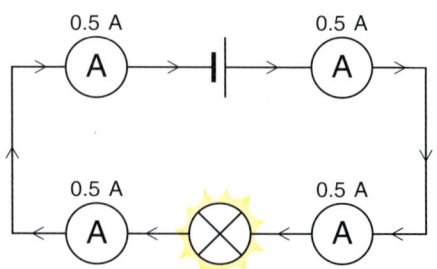

Figure 10 ▲ The current is the same all around a series circuit

In parallel circuits the current leaving a cell or battery is again the same as the current returning to it. The currents in the different branches of a parallel circuit, however, need not be the same.

Test Yourself

9 What is the name of the instrument used to measure the current flowing through a circuit?

10 In what units do we measure current?

Figure 11 ▲ The current need not be the same around a parallel circuit

Current and resistance

If we build the circuit shown in Figure 12 and close the switch, the bulb will glow brightly, showing that there is a good flow of charge around the circuit.

If a component called a **resistor** is included in the circuit, the bulb glows less brightly showing that the rate of flow of charge, i.e. the current, has decreased. The resistor is behaving as if it is an obstacle or constriction through which the charges have to travel. This hindrance to the flow of current is called **resistance**. Practically all components in a circuit have some resistance to the flow of current.

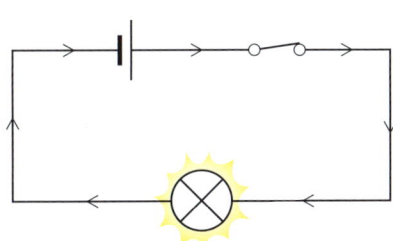

Figure 12 ▲

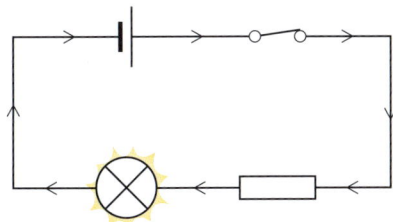

Figure 13 ▲

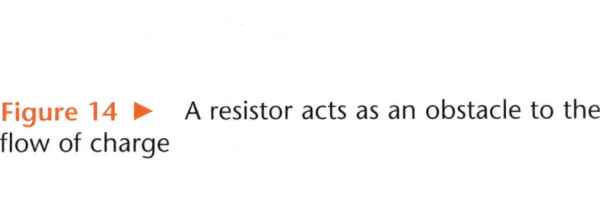

Figure 14 ▶ A resistor acts as an obstacle to the flow of charge

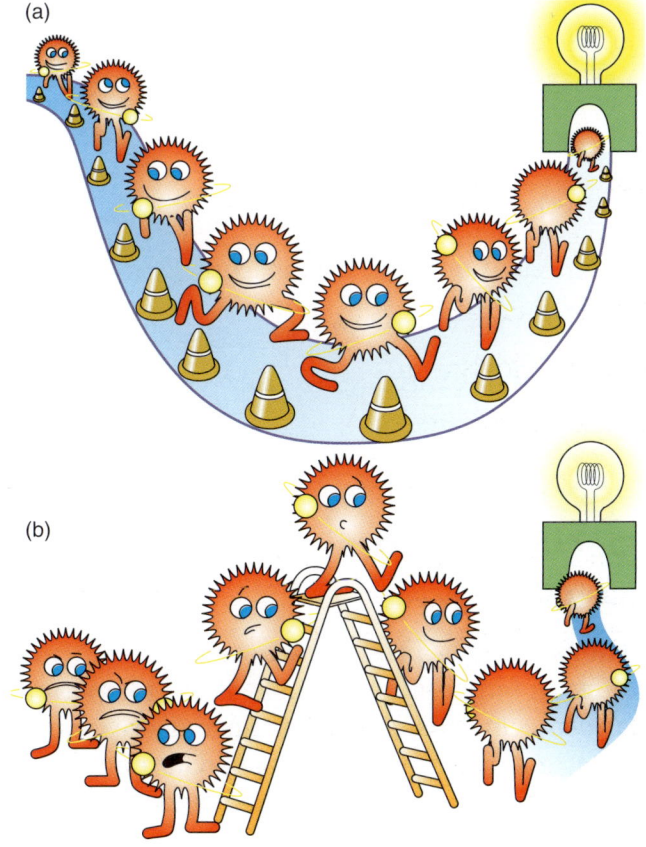

(a)

(b)

Resistance is measured in ohms (Ω). The larger the resistance of a component, the more difficult it is for current to flow through it. Connecting wires usually have resistances which are less than 1 Ω, whilst domestic light bulbs may have resistances as high as 1000 Ω.

Resistors are used in most circuits to control the size of the current that flows. Some resistors have fixed values, others have values that can be changed. They are called **variable resistors**.

By altering the value of the variable resistor in Figure 15, the bulb can be made to glow more or less brightly. This kind of circuit is often used in theatres and cinemas as well as in the home.

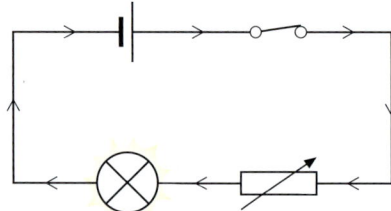

Figure 15 ▲ A dimmer circuit

Variable resistors have many uses. Each time you alter the brightness or contrast of your TV screens and computer monitors, you are adjusting a variable resistor. Even the loudness of your music systems or radios is controlled by a variable resistor

Test Yourself

11 In what units do we measure resistance?

12 What do we use resistors for in electrical circuits?

13 What is a variable resistor?

Special resistors

The thermistor

A **thermistor** is a resistor whose resistance alters a lot with change in temperature. Usually if the temperature of the thermistor increases, its resistance decreases. Thermistors are very useful in temperature dependant circuits such as might be found in fridges, freezers or fire alarms. The simple circuit in Figure 16 shows how they might be used.

As the temperature of the thermistor rises, its resistance decreases. The current flowing in the circuit increases and the indicator light glows brightly enough to warn of the temperature increase.

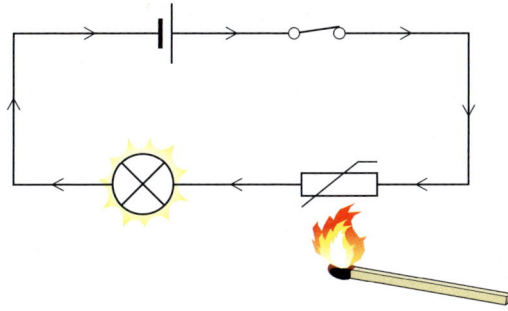

Figure 16 ▲   A circuit containing a thermistor

Light dependent resistor (LDR)

A **light dependent resistor** is a resistor whose resistance decreases when exposed to light. LDRs are used in light sensitive circuits such as automatic lighting controls for street lights and burglar alarms.

This simple burglar alarm warns if a light has been turned on. When the light shines on the LDR, its resistance decreases, a larger current flows around the circuit and the buzzer sounds.

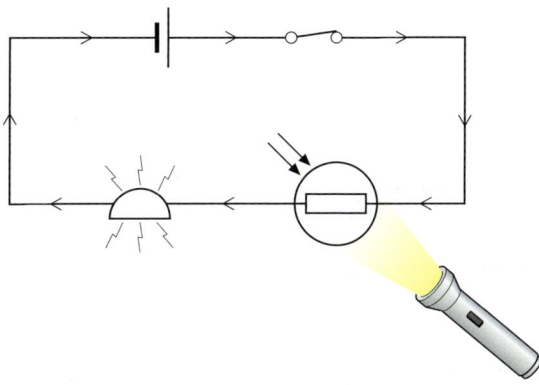

Figure 17 ▲   A simple alarm circuit

The diode

The **diode** is a resistor which behaves like a one-way street. When current flows though a diode in one direction, it can do so easily as the diode's resistance is low. But if current tries to flow in the opposite direction, the resistance of the diode is very high.

Diodes are therefore used in complicated circuits to prevent the flow of current in the 'wrong' direction.

A **light emitting diode** (LED) is a special kind of diode which emits light when current flows through it. The energy needed to make an LED glow is much less than that needed to make a bulb glow so LEDs are often used as indicator lights to show that electrical appliances are turned on or are in standby mode.

Figure 18 ▼   A diode allows the current to flow in one direction only

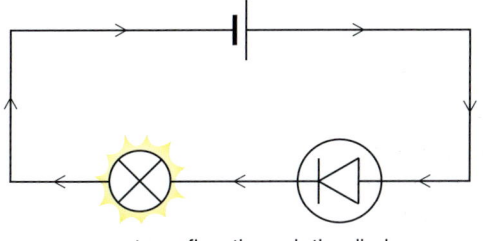

current can flow through the diode when it is connected like this ...

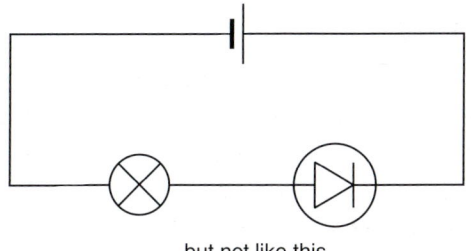

... but not like this

Voltages

As we have already seen, charges receive energy from cells or batteries. As they flow around a circuit this **electrical energy** is changed into other forms. For example when current passes through a bulb, some of the energy it is carrying is changed into heat and light energy. If current passes through a buzzer, some of the energy it is carrying is changed into sound.

We can measure how much energy charges are given as they pass through a cell or battery and how much of this electrical energy is transformed in the various components in a circuit by using a **voltmeter**. The voltmeter must be connected *in parallel* with the part of the circuit we are interested in.

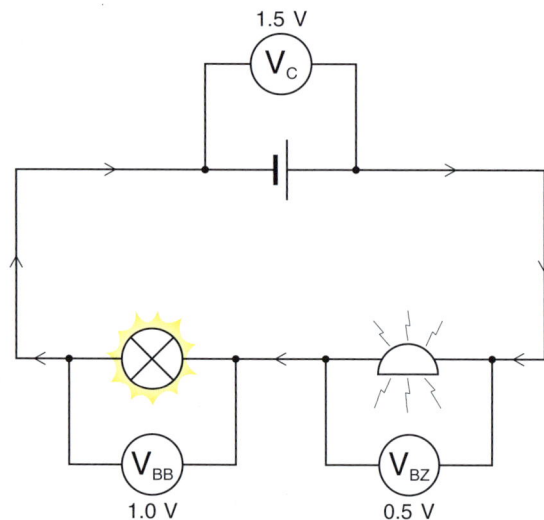

Figure 19 ▲ Measuring electrical energy in a circuit using a voltmeter

The voltmeter reading across the cell is equal to the sum of the voltmeter readings across the bulb and the buzzer. This indicates that all the energy received by the charges from the cell is converted into other forms of energy by the components in the circuit.

Test Yourself

14 Give one use for:
  a) a thermistor
  b) an LDR
  c) an LED.

Test Yourself

15 What does a voltmeter measure?

16 How should a voltmeter be connected into a circuit?

Summary

When you have finished studying this chapter, you should understand that:

✔ An electric current is a flow of charge.

✔ Conductors allow currents to pass through them easily.

✔ Insulators do not allow current to pass through them.

✔ Current will only flow if a circuit is complete.

✔ Series circuits have no junctions. Parallel circuit have junctions.

✔ We measure the size of a current by connecting an ammeter in series with the circuit. Current is measured in units called amperes or amps (A).

✔ All components in a circuit offer some resistance to the flow of current.

✔ Resistors are used to control the currents flowing in circuits. Special types of resistor include thermistors, LDRs and diodes.

✔ Currents carry electrical energy around circuits. This energy is changed into other forms of energy by the components in the circuit.

✔ We can measure electrical energy in a circuit using a voltmeter. A voltmeter must be connected in parallel with the component.

End-of-Chapter Questions

1 Explain in your own words the following key terms you have met in this chapter:

electric current
cell
battery
conductor
insulator
complete circuit
incomplete circuit
circuit diagram
component
series circuit
parallel circuit
conventional current

ampere
ammeter
resistor
resistance
variable resistor
thermistor
light dependent resistor
diode
light emitting diode
electrical energy
voltmeter

2 a) What do we call several cells connected together?

   b) Why is it important that the cells are connected the right way around?

   c) What is given to the charges as they pass through the cells?

   d) You are given three identical bulbs, some connecting wires and a cell. How many different circuits can you build using all these components? Draw a circuit diagram of each. In which circuit(s) will the bulbs glow brightest?

3 The diagram below shows two Christmas trees decorated with lights. Both trees have one bulb which is broken. None of the lights on tree A are glowing. All of the lights on tree B are glowing apart from the one that is broken.

End-of-Chapter Questions continued

a) What kind of lighting circuit does tree A have? Explain your answer.

b) What kind of lighting circuit does tree B have? Explain your answer.

4 Calculate the values of the currents flowing in the ammeters A–G in the circuits below.

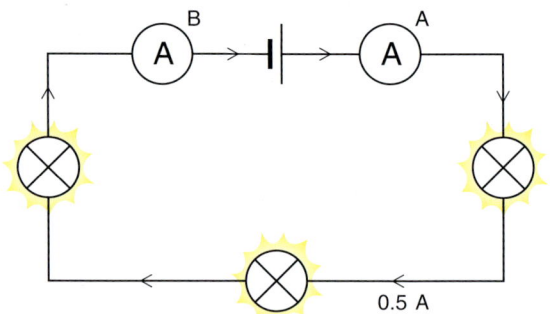

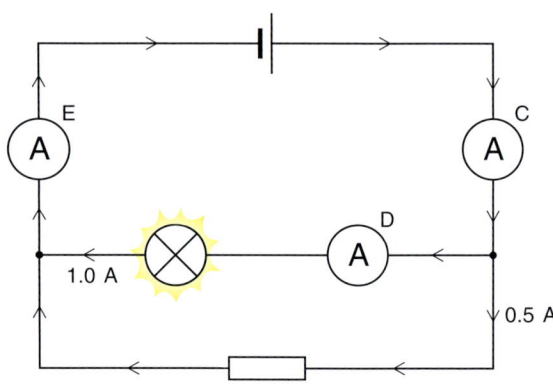

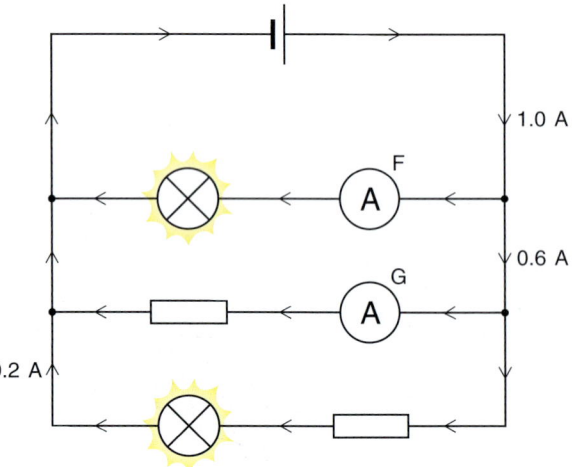

5 Look at the circuit below

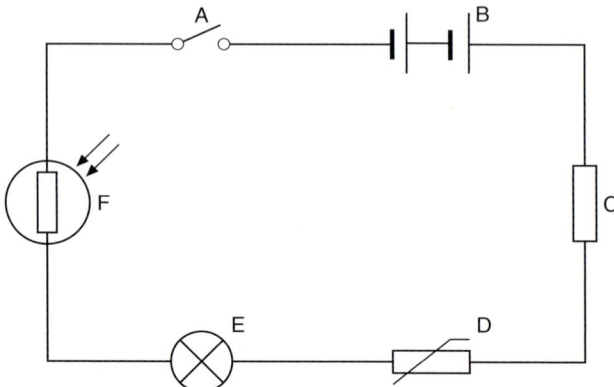

a) Identify all the components A to F.

b) If component A is closed and component D is warmed, explain what would happen to component E.

c) Give one use for component D.

d) Give one use for component F.

6 Look carefully at the circuit drawn below:

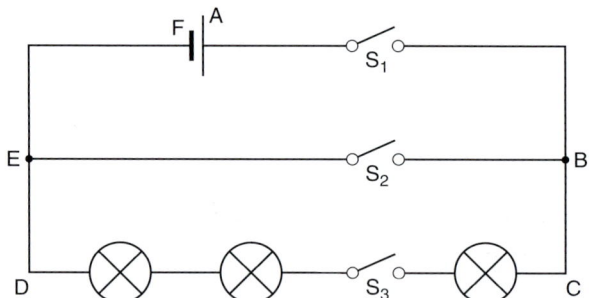

a) If all the switches in this circuit are closed, which path offers the least resistance to the current?

b) What will happen to all the bulbs in the circuit when all the switches are closed?

c) Why is this kind of connection called a **short circuit**?

d) Why might a short circuit damage a cell or battery?

4 Magnetism

It really doesn't matter how long this student tries to pick up the pieces of paper with her magnet; she is not going to succeed. Magnets will only attract objects that are made from **magnetic materials**. Examples of magnetic materials include iron, steel, nickel and cobalt. Paper is a **non-magnetic material**. It is not attracted by a magnet. Other examples of non-magnetic materials include wood, plastic, aluminium and copper.

If we sprinkle some iron filings onto a bar magnet or horseshoe magnet, we see that most of the filings are attracted to the ends. It is here where the magnet is strongest. These are called the **poles** of the magnet.

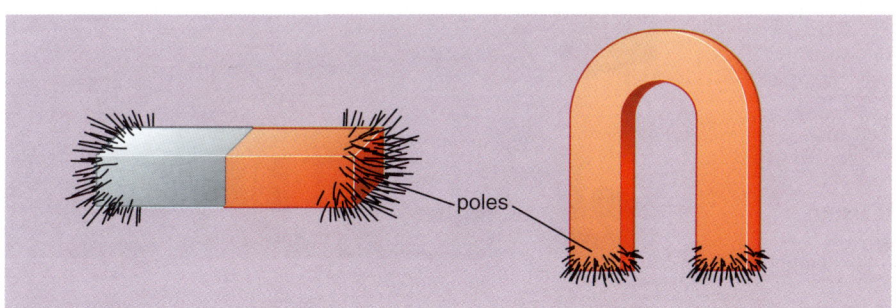

poles

Figure 1 ▲   The poles are the strongest parts of a magnet

Most magnets have two poles, a **north pole** and a **south pole**. The simple experiment described below will tell you how the poles got these names

If a bar magnet is suspended in a paper stirrup, as shown, providing there are no other magnets or magnetic materials close by, the magnet will come to rest lying in a North–South direction.

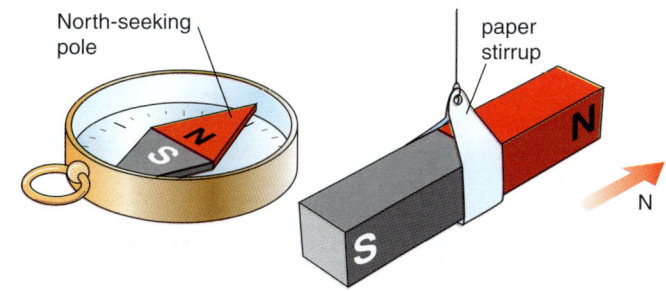

North-seeking pole

paper stirrup

N

Figure 2 ▲   How the north and south poles of a magnet got their names

The end of the magnet that is pointing north is called the north-seeking pole, now shortened to the north pole. The end that is pointing south is called the south-seeking pole, shortened to the south pole. The bar magnet is in fact behaving like a **compass**.

Test Yourself

1   What is a magnetic material? Give two examples of magnetic materials.

2   What is a non-magnetic material? Give two examples of non-magnetic materials.

3   Explain why one end of a bar magnet is called the south pole.

Attraction and repulsion

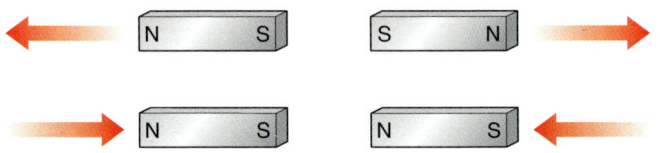

Figure 3 ▲   Like poles repel, while unlike poles attract

Figure 4 ▶   This is the Maglev train. It uses magnetic repulsion to levitate above the rails

If two similar poles are placed near each other they repel.

If two dissimilar poles are placed near each other they attract.

The domain theory of magnetism

No one completely understands all the mysteries of magnetism but most of the basic ideas can be explained using the 'domain theory'. There are several kinds of magnetism including diamagnetism, paramagnetism and ferromagnetism. The one we shall be investigating is ferromagnetism.

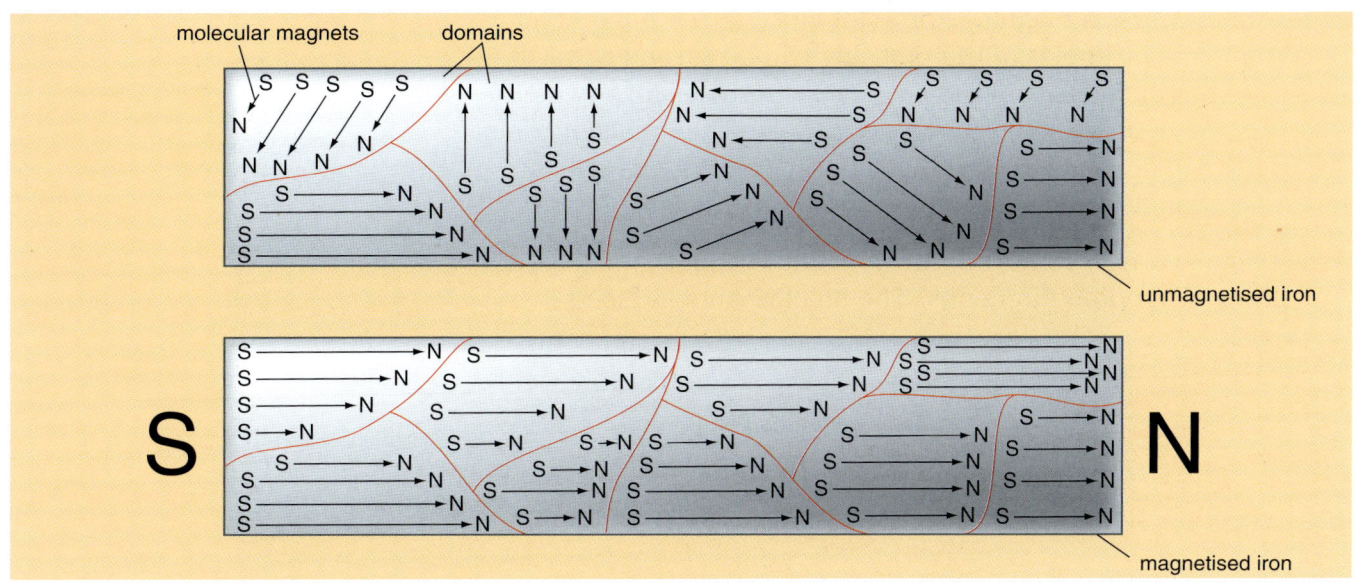

Figure 5 ▲   The domains in a piece of magnetised and unmagnetised iron

We believe that ferromagnetic materials, such as iron and steel, have inside them small **molecular magnets**. These small molecular magnets are contained in tiny cells called **domains**. Within each domain all the molecular magnets point in the same direction.

In an unmagnetised piece of iron, the molecular magnets in each domain point in a different direction. As a result there is no overall magnetic effect, neighbouring domains cancel each other out.

Within a magnet, the molecular magnets within neighbouring domains all point in the same direction. As a result the magnetic effects of the domains reinforce each other, producing a magnet. A material in which the domains are not aligned is described as being unmagnetised. A material in which the domains are aligned is described as being magnetised.

Using the domain theory to explain why things happen

What happens to a magnet if it is cut in half? What happens to a magnet if it is continually knocked or dropped? Why does a magnet attract a piece of unmagnetised iron?

If a magnet is cut or broken in half, it will produce two smaller magnets because the domains in each half are still aligned.

If a magnet is continually dropped or hit it will lose its magnetism as the domains will be knocked out of line.

Test Yourself

4   Explain how a Maglev train is able to hover above the rails.

5   Draw a diagram to show the domains in a bar of iron that is a) unmagnetised and b) magnetised.

If the north pole of a magnet is brought close to the head of an iron nail, the south poles of the domains will be turned towards the head creating a south pole here. Opposite poles attract so the head of the nail is attracted to the north pole of the magnet. This aligning of the domains in an unmagnetised material when a magnet is brought near is called **induced magnetism**.

If the magnet is now removed, the domains within the nail return to pointing in random directions. The nail loses its induced magnetism.

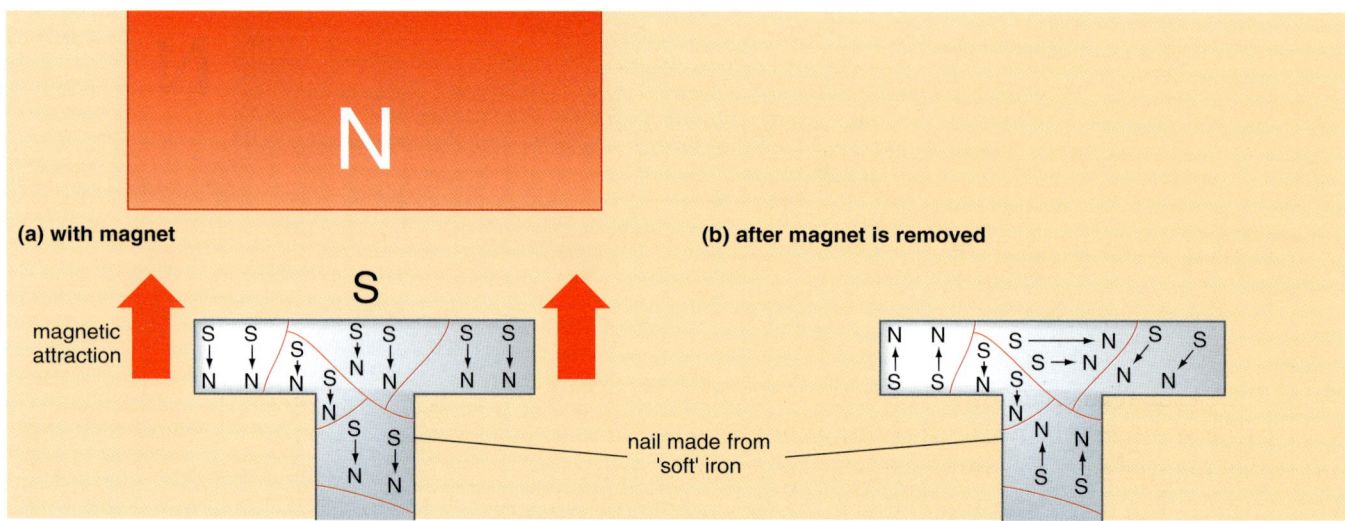

If the experiment is repeated with a steel nail this is not the case. When the magnet is moved away from the nail many of the domains remain aligned. The nail has retained its magnetism

Figure 6 ▲ This nail is made of iron – a magnetically soft material

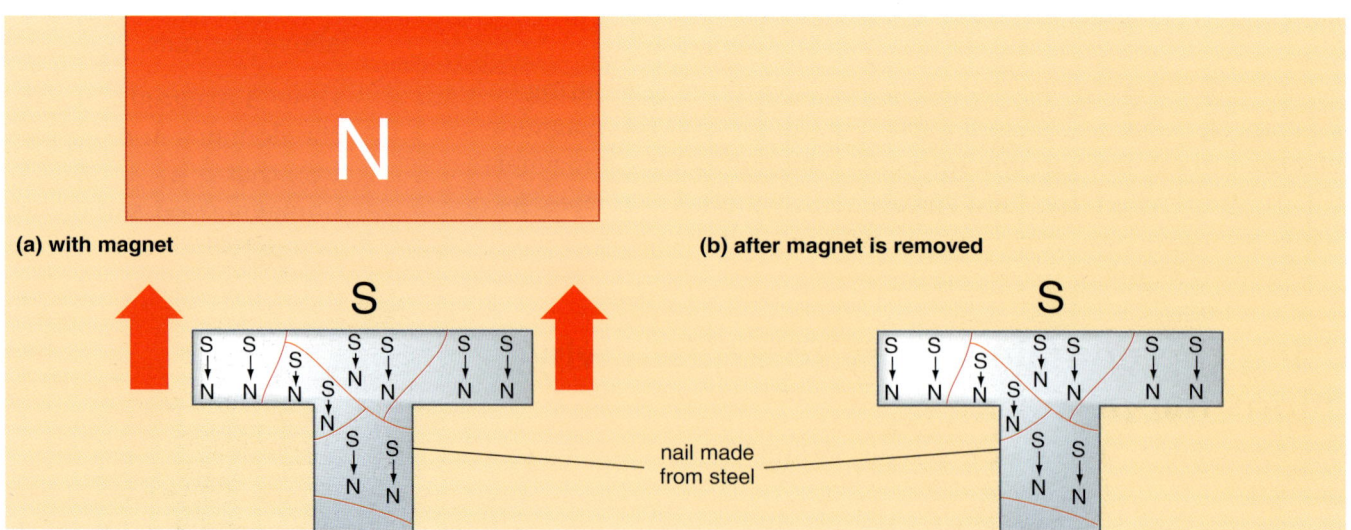

Materials which easily gain and lose their magnetism are called **magnetically soft** materials. Iron is a magnetically soft material. Materials that are a little more difficult to magnetise but whose domains remain aligned when the magnet is removed are called **magnetically hard** materials. Steel is a magnetically hard material.

Figure 7 ▲ This nail is made of steel – a magnetically hard material

Reed switches

A reed switch consists of a small glass tube containing two soft iron reeds. When there is no magnet close by, there is a gap between the reeds i.e. the switch is open. If a magnet is brought close to the switch, the reeds become magnetised, attract each other and the switch closes.

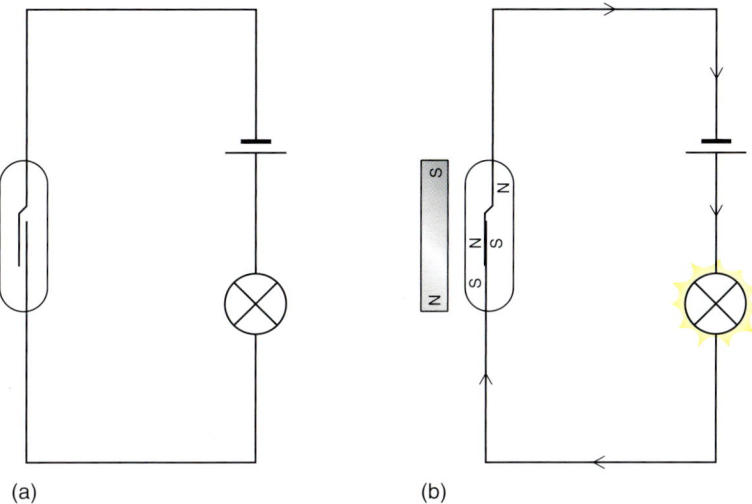

(a)   (b)

Figure 8 ▲   A reed switch

Some reed switches are designed to work the opposite way around, i.e. they are open when there is a magnet close by and they are closed if the magnet is removed. Figure 9 shows how such a switch could be used in a burglar alarm.

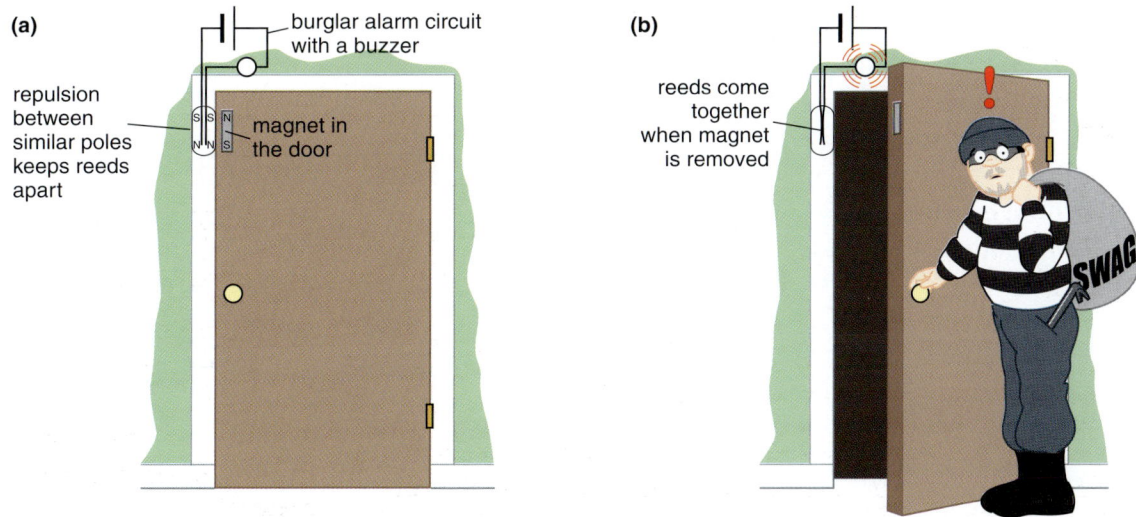

Figure 9 ▲   With the door closed, this reed switch is open. But when the burglar opens the door, the switch closes and the alarm sounds

Making magnets

There are several ways in which a magnetic material can be magnetised.

1 The single touch method

If a steel rod is stroked 15–20 times with one end of a bar magnet, the domains can be made to line up so that they are all pointing in the same direction. The rod is then magnetised. Because the rod is made from a magnetically hard material, it remains magnetised.

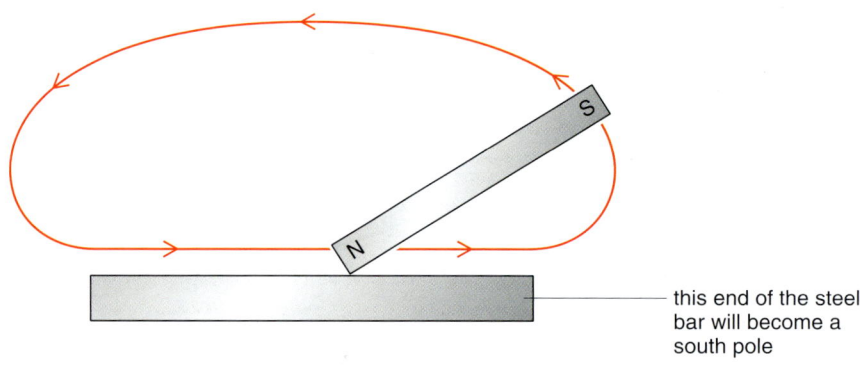

this end of the steel bar will become a south pole

Figure 10 ▲ The single touch method of making a magnet

2 The electrical method

When an electric current flows through a long coil or solenoid, it creates a strong magnetic field. If a steel rod is placed through the centre of the coils this magnetic field can be used to line up all the domains and magnetise the rod.

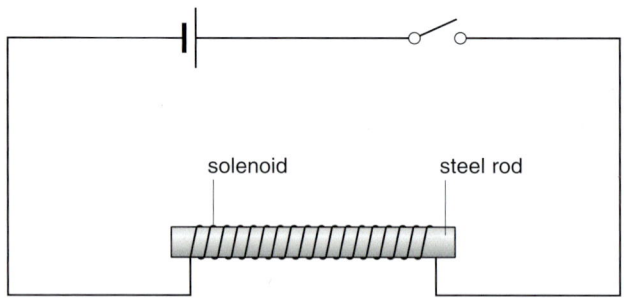

solenoid   steel rod

Figure 11 ▲ Using a solenoid to magnetise a steel bar

Magnetic fields

When an object made of iron or steel is placed close to a magnet, it is attracted towards the magnet even though it is not in contact. This happens because the object is inside the **magnetic field** of the magnet. A magnetic field is any volume of space where magnetism can be detected.

Discovering the shape of the magnetic field around a bar magnet using iron filings

A piece of paper is placed over the magnet and iron filings gently sprinkled over the paper. The pattern formed by the filings shows the shape of the magnetic field.

Discovering the shape of the magnetic field around a bar magnet using a small compass

A magnet is placed on a piece of paper and drawn around. A small plotting compass is then placed next to the magnet, a circle drawn around it and the direction in which the compass points marked inside the circle. The compass is then moved so that its tail is next to the compass point just drawn. A circle is again drawn around the compass and its direction recorded. This process is repeated all over the paper. The pattern of compass needles on the paper show the shape and direction of the magnetic field.

We usually draw a magnetic field using **field lines**. These lines tell us several things about a magnetic field.

- They show the *shape* of a magnetic field.

- They show the *direction* of a magnetic field. Magnetic field lines always 'travel' from north to south.

- They show the *strength* of a magnetic field at any particular place. If the lines are close together, this indicates that the field is strong here. If the lines are very spread out, this indicates that the field is weak here.

Note: Magnetic field lines *never* touch or cross.

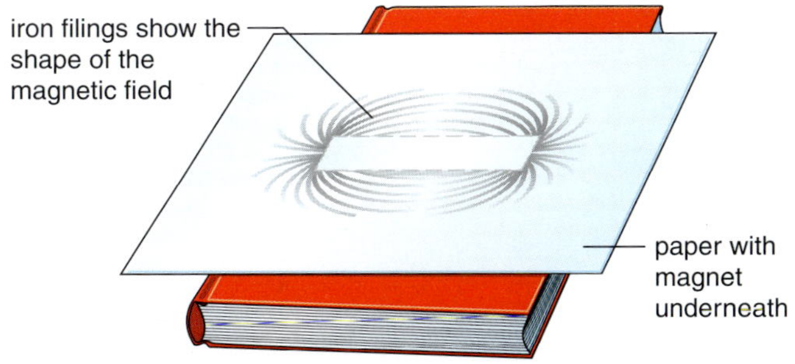

iron filings show the shape of the magnetic field

paper with magnet underneath

Figure 12 ▲ Iron filings can be used to show the shape of the magnetic field around a magnet

Figure 13 ▼ The magnetic field around a magnet can also be shown using a plotting compass

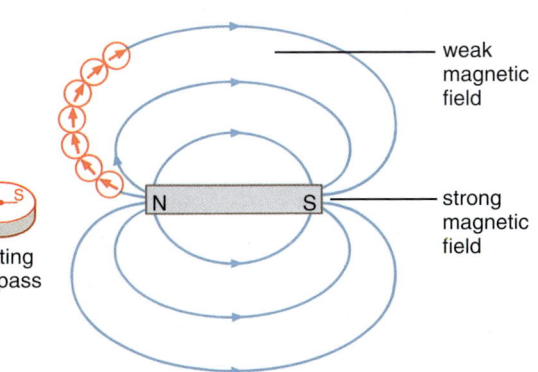

weak magnetic field

strong magnetic field

plotting compass

Extension box

Overlapping magnetic fields

If two or more magnets are placed close to each other, their magnetic fields will overlap. In those places where the fields are in the same direction they will combine to produce a stronger field. In those places where the fields oppose each other they will combine to produce a weaker magnetic field.

In some places the overlapping magnetic fields may overlap and cancel each other out i.e. there is no magnetic field at these places. They are known as **neutral points**.

Magnetic shielding

A magnet held below a table can be used to move a steel toy car on top of the table, creating the illusion that the car is moving unaided. The magnet is able to move the car because its magnetic field can pass through the wooden table. If, however, a thin sheet of iron is placed under the car, the magnet is no longer able to do this.

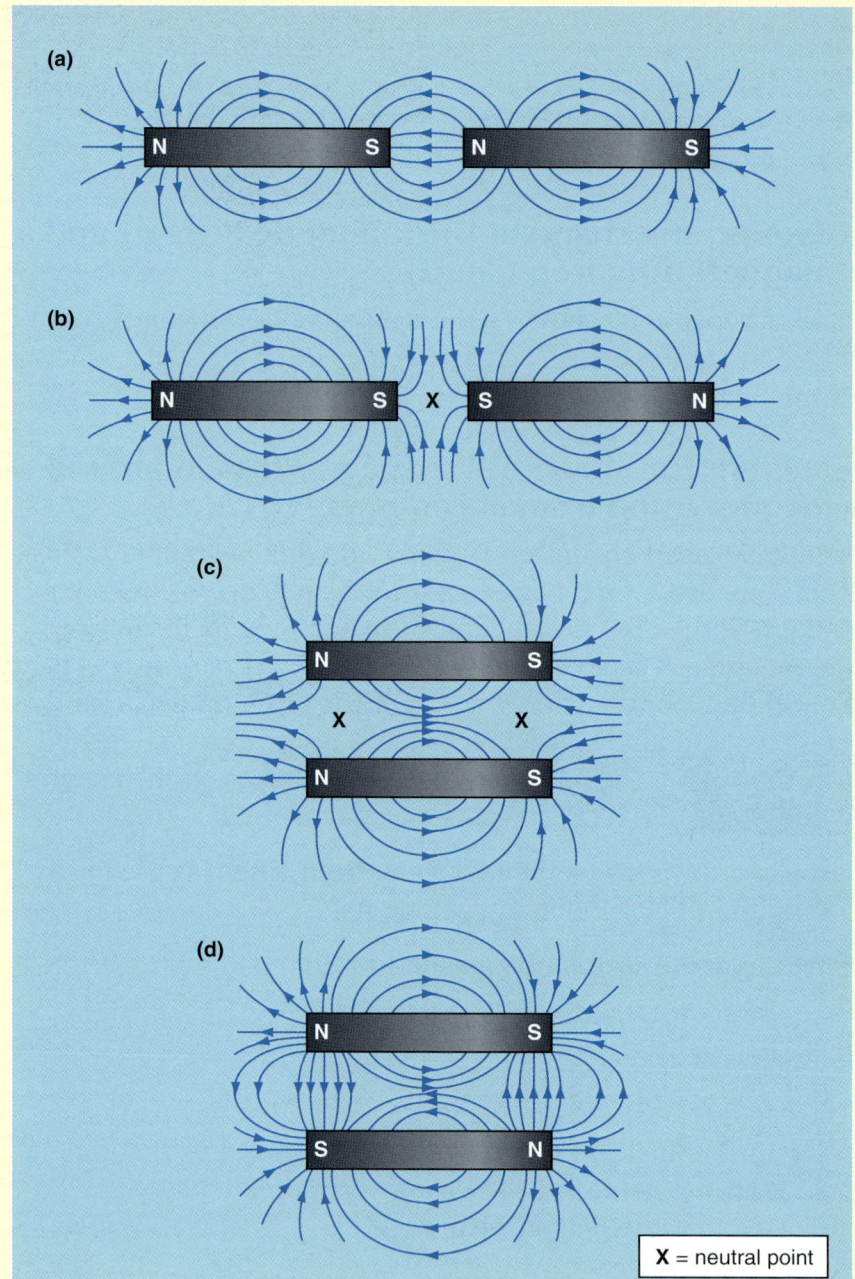

(a)

(b)

(c)

(d)

X = neutral point

Figure 14 ▲ Shapes of magnetic fields

A magnet's magnetism is able to pass through a non-magnetic material but it cannot pass through a magnetic material. Magnetic materials can therefore be used to shield objects and apparatus such as sensitive electrical circuits from stray magnetic fields.

Extension box continued

The Earth's magnetic field

The Earth is surrounded by a magnetic field. Its shape and direction is as if there is a giant magnet buried within the Earth lying almost parallel to the axis of rotation.

It is this magnetic field which for centuries has allowed travellers to navigate using a compass. It is thought that the field is created by large currents deep inside the Earth's core.

Find out

Why are the magnetic north pole and the true North Pole not in the same place?

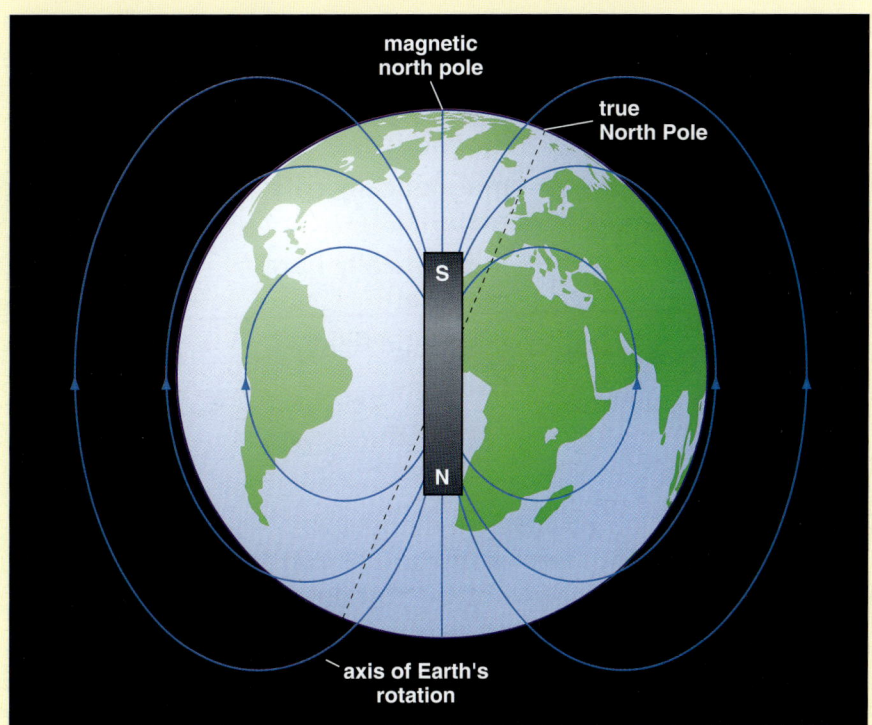

Figure 15 ▲   The Earth has a magnetic field

Summary

When you have finished studying this chapter, you should understand that:

✔ Magnets are able to attract objects made from magnetic materials such as iron and steel.

✔ The strongest parts of a magnet are called its poles. Most magnets have two poles, a north pole and a south pole.

✔ Opposite poles attract. Similar poles repel.

✔ Magnetic materials contain small cells called domains. Within each domain are mini magnets that all point in the same direction.

✔ In an unmagnetised piece of iron or steel, the mini magnets in neighbouring domains point in different directions.

✔ In a magnetised piece of iron or steel, the mini magnets in neighbouring domains point in the same direction.

✔ Magnetically soft materials such as iron easily lose their magnetism. Magnetically hard materials such as steel hold onto their magnetism more strongly.

✔ Around a magnet is a volume of space called a magnetic field where magnetism can be detected.

✔ Magnetic field lines are used to show the shape, strength and direction of a magnetic field.

End-of-Chapter Questions

1 Explain in your own words the following key terms you have met in this chapter:

magnetic material

non-magnetic material

pole of a magnet

north pole

south pole

compass

molecular magnet

domain

induced magnetism

magnetically soft material

magnetically hard material

magnetic field

field lines

neutral point

2 Explain how you could use a magnet to remove iron filings from a mixture of iron filings and sawdust. Why could you not use the same procedure to separate a mixture of copper filings and sawdust?

3 Draw a diagram of the magnetic field around a bar magnet. Mark on your diagram a) a place where the magnetic field is strong and b) a place where the magnetic field is weak. Explain how your diagram shows the strength of the magnetic field at different places.

4 The diagram below shows the 'indian rope trick' done by using a magnet. The attraction between the magnet and the paper clip holds the thread taut.

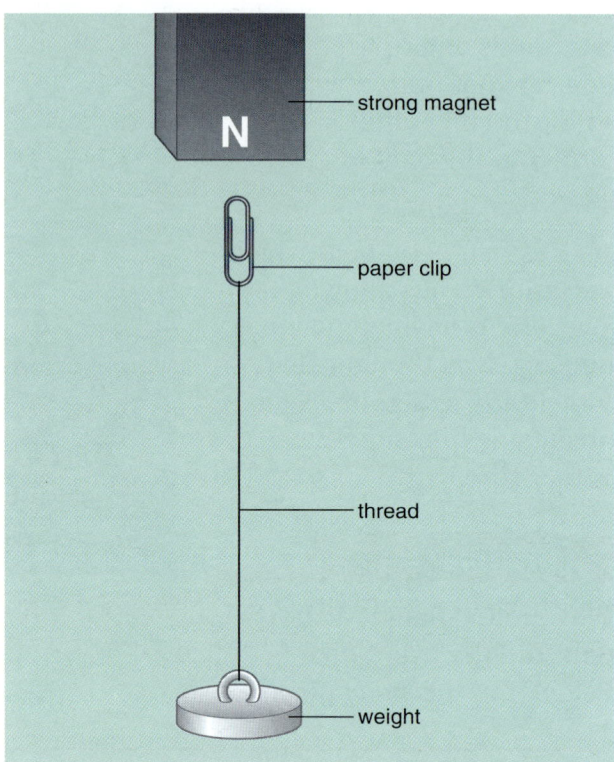

Explain what happens when a) a piece of card and b) a thin sheet of steel is placed between the magnet and the paper clip.

5 A pupil is given five identical steel bars labelled A, B, C, D and E. She is told that at least two of the bars are magnetised. Write down a set of instructions the pupil should follow so that she can discover how many of the bars are magnetised and which they are. No other apparatus is available.

6 Why does an isolated compass always point in a N–S direction. Why does the compass have to be isolated? Find out how ships isolate their compasses.

5 Electromagnetism

In 1819, a Danish scientist named Oersted noticed that when he passed a current through a wire, a compass nearby moved so that it was pointing in a new direction.

When he turned the current off, the compass returned to pointing in a N–S direction. Oersted realised that when currents flow through wires, magnetic fields are created. Oersted had discovered **electromagnetism**.

The shape and direction of these magnetic fields can be seen using iron filings and compasses.

When the switch is closed, current flows through the wire AB. The four compasses move from a N–S direction to new positions showing that there is a magnetic field. The directions in which the compasses are now pointing suggest that the field around the wire is circular. If iron filings are sprinkled onto the card, they form a circular pattern.

If the switch is opened, no current flows through the wire AB, the compasses return to pointing northwards and any new iron filings sprinkled onto the card form no pattern.

From these simple experiments we can conclude that when a current flows through a wire, a circular magnetic field is created around it. If there is no current, there is no magnetic field.

If the direction of the current in AB is reversed, the direction of the magnetic field also changes.

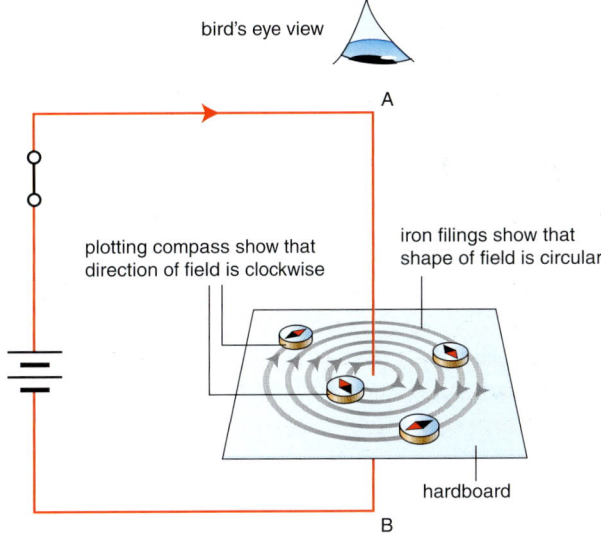

Figure 1 ▲ When a current is flowing through a wire, a magnetic field is created

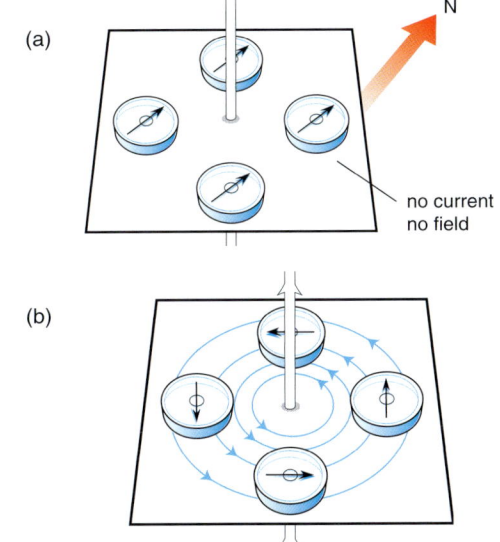

Figure 2 ▶ a) When no current flows through the wire, there is no circular magnetic field. b) The direction of the magnetic field depends on the direction of the current

Extension box

We can predict the direction of the magnetic field created around a wire when a current flows along it using Maxwell's corkscrew rule.

Imagine you are holding a corkscrew or screwdriver so that it is parallel to the wire. Now turn the corkscrew or screwdriver so that

the screw travels in the same direction as the current in the wire. If you have turned the corkscrew or screwdriver clockwise, then the magnetic field around the wire will also be clockwise. If your movement is anticlockwise, the magnetic field is also anticlockwise.

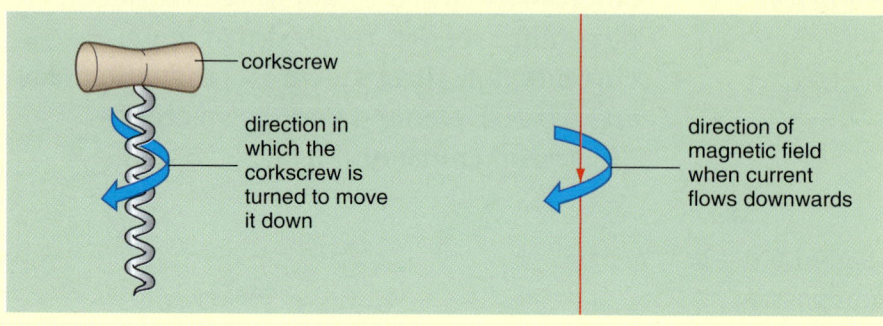

corkscrew

direction in which the corkscrew is turned to move it down

direction of magnetic field when current flows downwards

Figure 3 ▲ Maxwell's corkscrew rule

Test Yourself

1   What is the shape of the magnetic field around a wire through which a current is flowing?

2   Describe two ways in which you could prove a) that there is a magnetic field around the wire and b) that the field is only present when current flows through the wire.

3   What happens to the magnetic field around a wire if the direction of the current is reversed?

Coils and solenoids

The magnetic field created around a wire when a current flows through it is quite weak. If, however, the wire is wrapped into several loops, the fields around each of the pieces of wire overlap, creating a stronger magnetic field. If the wire is wrapped to form a long **coil**, this is called a **solenoid**. The field has the same shape as the magnetic field around a bar magnet.

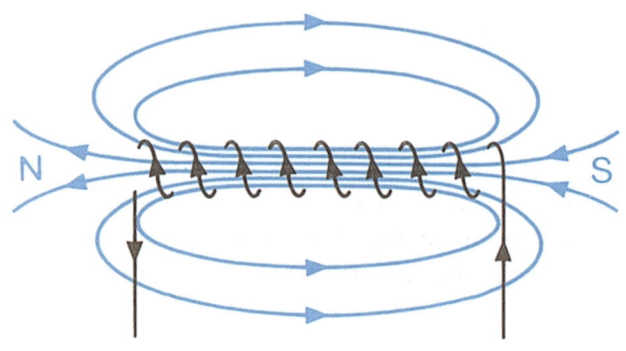

Figure 4 ▲   The magnetic field around a solenoid is the same shape as that around a bar magnet

We can predict which end of a coil has a north pole and which end has a south pole using the right hand grip rule. Imagine wrapping the fingers of your right hand around a solenoid so that they are pointing in the direction the current is flowing through the coils. Your thumb will now be pointing towards the north pole of the solenoid.

If the direction of the current through the solenoid is changed, the polarity of the field also changes.

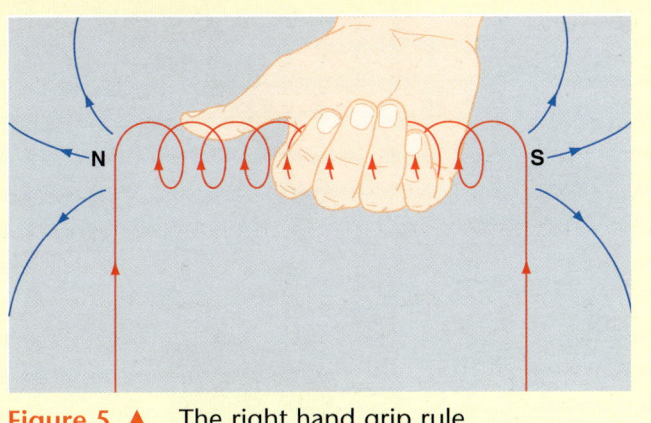

Figure 5 ▲ The right hand grip rule

Electromagnets

If a piece of soft iron is placed down the centre of a solenoid, the strength of the magnetic field increases. This combination of solenoid and **soft iron core** is called an **electromagnet**.

Electromagnets are extremely useful because

- they can be turned on and off
- their magnetic strengths can be changed.

Turning electromagnets on and off

When the switch S is closed, current flows through the solenoid. The soft iron core becomes magnetised and attracts objects made from magnetic materials.

When the switch is opened, no current flows through the solenoid. The soft iron core loses its magnetism and the objects are no longer attracted.

If the core of an electromagnet was made from a magnetically hard material such as steel, it would keep some of its magnetism even when the current was turned off.

Figure 8 shows an electromagnet being used in a scrapyard. When the current is turned on, the electromagnet is able to pick up objects that contain iron or steel. When objects have been moved to where they are required, the current is turned off, the electromagnet loses its magnetism and the objects fall.

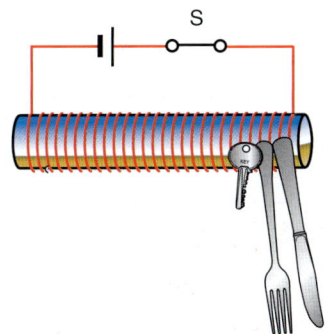

Figure 6 ▲ An electromagnet attracts magnetic objects when switched on

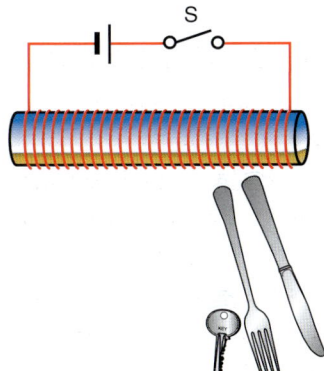

Figure 7 ▲ When the switch is opened, the magnetism is lost and the objects fall

Figure 8 ◄ One use of an electromagnet

Altering the strength of an electromagnet

There are two ways in which we can increase the strength of an electromagnet:

1 increase the number of turns on the solenoid
2 increase the current flowing through the solenoid

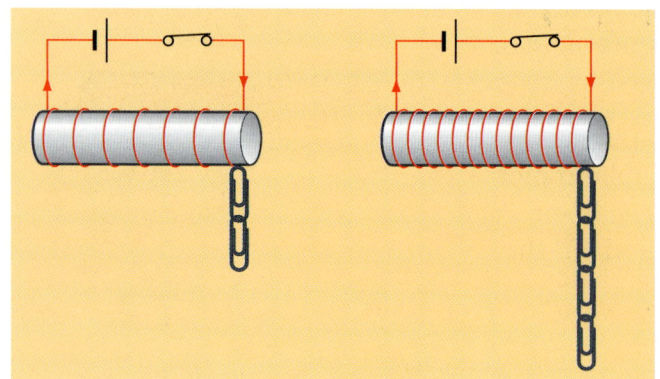

Figure 9 ▲ The electromagnet is stronger when the coil has more turns

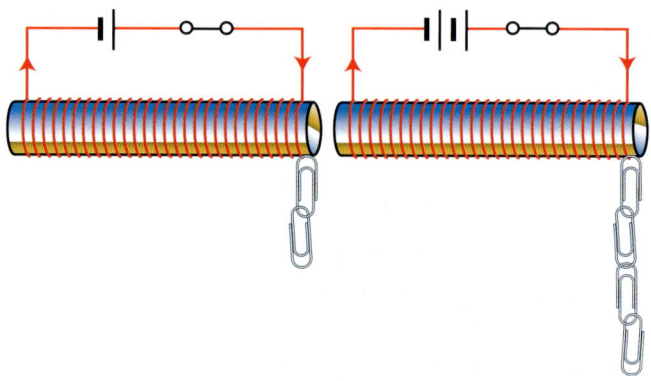

Figure 10 ▲ The electromagnet is stronger when there is a larger current flowing through it

The electromagnet circuit breaker

Figure 11 shows a trip switch or **circuit breaker**. This is used to limit the size of the current flowing along a wire.

If the current flowing through the circuit breaker becomes too large, the electromagnet E becomes strong enough to pull the iron catch to the right. This allows the contacts at C to part, making the circuit incomplete. When the fault that caused too high a current to flow has been corrected, the reset switch is pressed and the circuit is again complete.

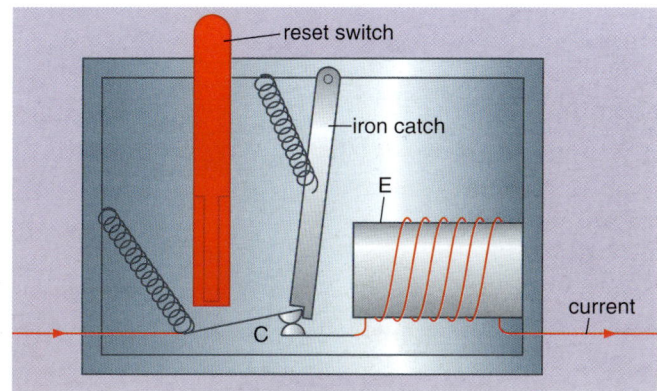

Figure 11 ▲ A circuit breaker

The electric bell

When the bell push is pressed, current flows around the circuit and the electromagnet E becomes magnetised. The iron rod is attracted towards it and the hammer hits the bell. While this is happening, a gap appears at the contact screw. The circuit is now incomplete, the electromagnet loses its magnetism and the spring pulls the rod back to its original position. The whole process then begins again. The bell continues to ring as long as the bell push is pressed.

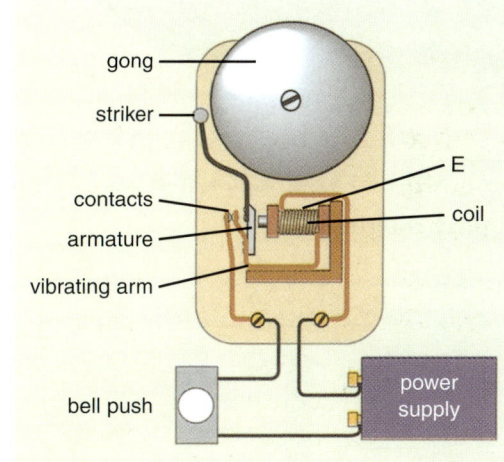

Figure 12 ▲ An electric bell

The relay switch

Sometimes it is useful to be able to control the current flowing in one circuit by using a second circuit. This is especially true if the current flowing in the second circuit is large and therefore may be dangerous. This can be done using a device called a **relay switch**.

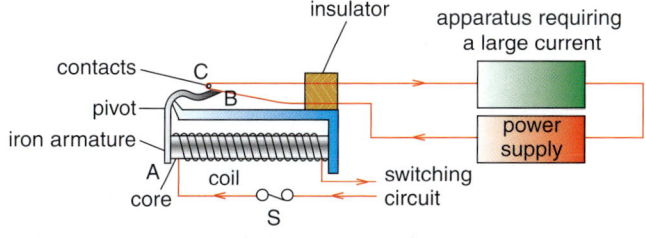

Figure 13 ▲ A relay switch

When the switch S is closed, the iron core becomes magnetised and attracts the iron armature. As one end of the armature A is attracted, its other end B pushes the contacts at C together. The second circuit is now complete and therefore turned on. If the switch S is opened, the electromagnet loses its magnetism, the armature returns to its original position and the second circuit is turned off.

The electric motor

Figure 14 shows the French train 'the TGV'. Its powerful electric motors enable it to reach speeds in excess of 300 km per hour.

Figure 14 ◄ The TGV uses powerful electric motors

Electric motors change electrical energy into kinetic energy using overlapping magnetic fields.

Figure 15 ▼

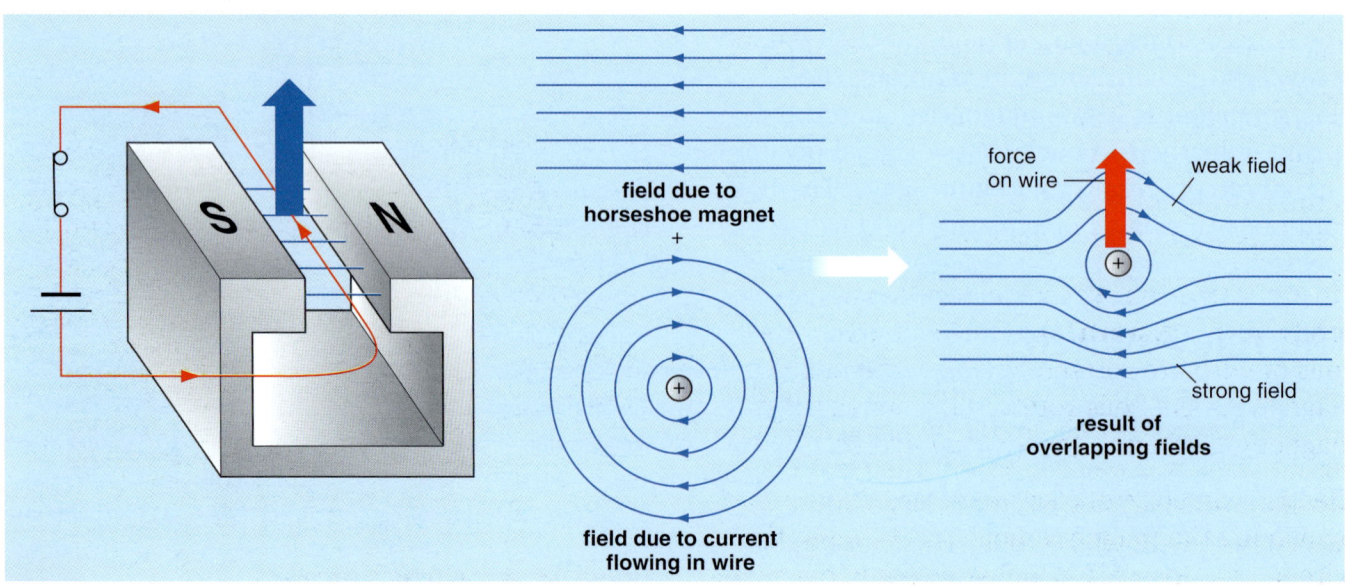

field due to horseshoe magnet

+

field due to current flowing in wire

force on wire

weak field

strong field

result of overlapping fields

If current is passed through a wire placed between the poles of a magnet, the wire will be seen to move. This happens because the magnetic field around the wire and the magnetic field across the poles of the magnet overlap. In those places where the fields are in the same direction, they reinforce each other. In those places where they are in opposite directions, they produce a weaker magnetic field. The wire then experiences a force moving it from the stronger part of the field into the weaker part. It is this force which causes the wire to move. If the direction of the current in the wire or the magnetic field is changed, the direction of the force on the wire also changes.

If the wire is made into a loop, the currents flowing through opposite sides will be in different directions. As a result, the forces created will cause the loop to rotate.

In real motors, the single loop of wire is replaced with a coil of wire with many turns and the permanent magnet is often replaced with an electromagnet.

Generating electricity

We all use electricity. Without it, life would be very different. Televisions, microwaves and fridges would not work if we were unable to generate electricity.

If the wire in Figure 17 is moved up or down so that it cuts through the field lines of the magnet, the needle of the ammeter is seen to move, showing that current is flowing through the wire. When the wire is stationary, no current flows. It is the movement of a wire through a magnetic field which we use to generate electricity.

Most of the electricity we use in our homes is produced by **generators**. Inside a generator, coils of wire are rotated between the poles of a magnet. As the coils rotate, they cut through magnetic field lines generating current.

Electric current can also be generated in wires by moving the magnetic field and keeping the wire stationary. This principle is used in a simple bicycle dynamo.

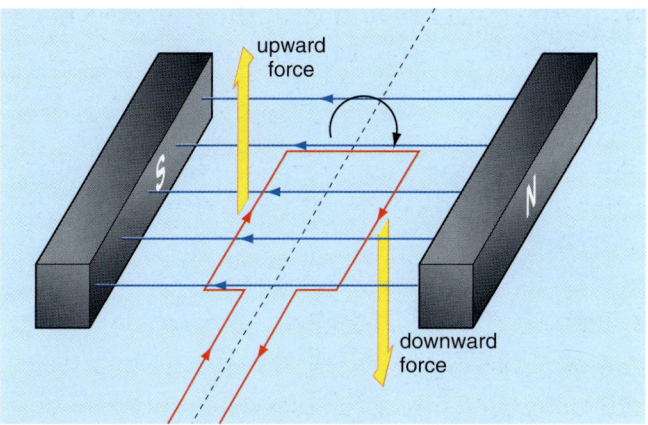

Figure 16 ▲   A loop of wire in a magnetic field

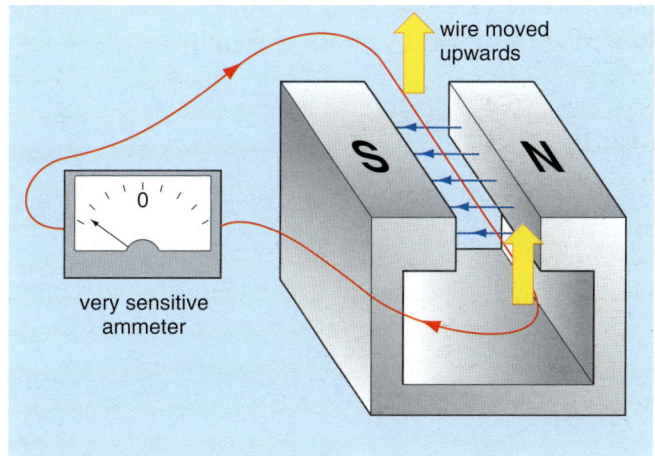

Figure 17 ▲

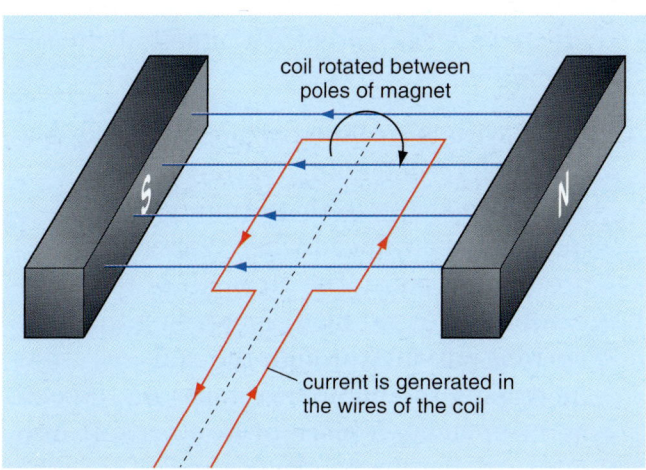

Figure 18 ▲   A simple generator

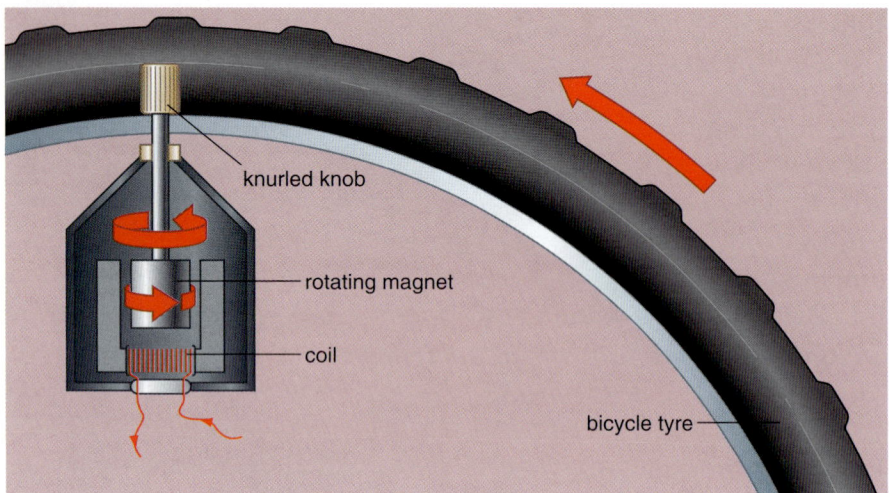

Figure 19 ◄ How a bicycle dynamo works

The turning wheel causes the knurled knob to turn. The axle and the magnet attached to it rotate. The magnet's magnetic field cuts through the wires of a coil. Current is generated in the wires, which is then used to light the bulb.

Test Yourself

7 Draw a diagram of the magnetic field around a single piece of wire which has a current passing through it.

8 Draw a diagram of the magnetic field between the poles of a horseshoe magnet.

9 Draw a diagram of the field produced when the above two fields overlap.

Summary

When you have finished studying this chapter, you should understand that:

✔ When a current flows through a wire, it creates a weak, circular magnetic field around it.

✔ If the wire is wrapped around to make a long coil or solenoid, a stronger field, similar in shape to that around a bar magnet, is produced when current flows around the coils. The strength of this field can be increased by a) increasing the current passing through the coils, b) increasing the number of turns on the coil and c) placing a piece of iron – a soft iron core – down the centre of the coil.

✔ The combination of the soft iron core and the coil is called an electromagnet.

✔ Electromagnets are used in many devices including electric bells, relay switches and electric motors.

✔ If a wire cuts through a magnetic field or a magnetic field moves across a wire, current can be 'generated'.

End-of-Chapter Questions

1 Explain in your own words the following key terms you have met in this chapter:

electromagnetism	electromagnet
coil	circuit breaker
solenoid	relay switch
soft iron core	generator

2 a) Draw a diagram showing the magnetic field around an electromagnet when it is turned on.

b) Suggest two ways in which you could show the shape of the field.

c) Explain what would happen if the electromagnet used in a scrapyard had a core made from a magnetically hard material such as steel.

3 a) Name three devices that use electromagnets.

b) Draw a diagram and explain how one of these devices works.

4 The diagram below shows the circuits used to start a car engine.

a) Explain how turning the key makes the starter work.

b) Give one reason why a relay switch like this is used when starting a car.

5 a) Suggest three ways in which the rate of rotation of an electric motor can be increased.

b) Suggest two ways in which the direction of rotation of an electric motor can be changed.

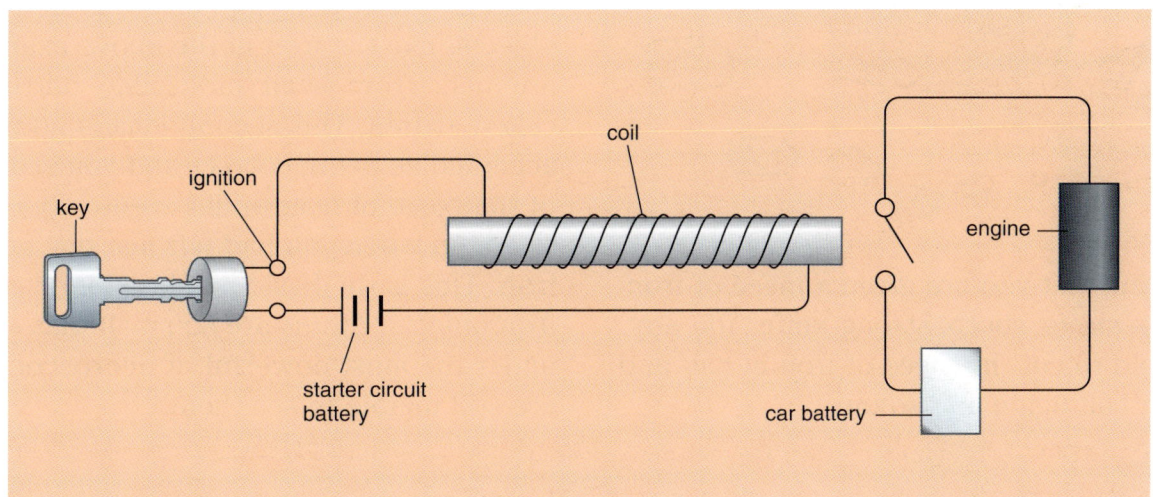

6 A reed switch is opened and closed using a magnet. A reed relay switch uses the magnetic field created by a coil.

Explain what happens when switch S is closed. Give one use for a reed relay switch.

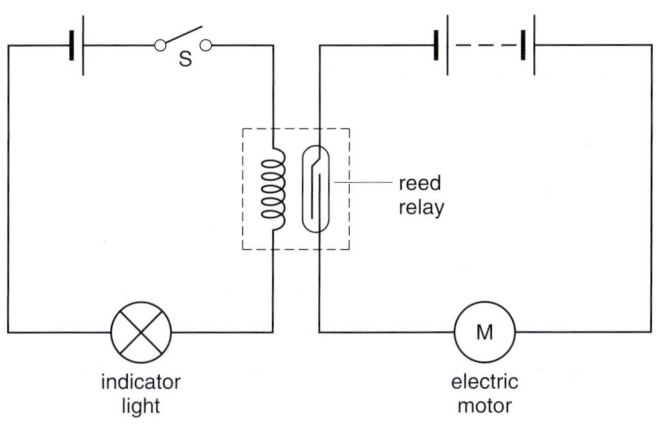

6 Heating and cooling

The building blocks of matter

Imagine that you are given a piece of iron and a *super* sharp, strong knife. You are then told to cut the piece of iron in half and then in half again and so on.

Assuming that you could see what you were doing and the blade of the knife was infinitely sharp, could you carry on doing this forever or would there come a point where you would be unable to cut the iron in half? This problem has been debated by philosophers for over 2000 years. The Greek philosopher Aristotle suggested that matter was continuous and you would theoretically be able to cut a piece of matter into smaller pieces without any limit. Democritus, another Greek philosopher, suggested that objects, like the iron in the above example, are made from small 'indivisible' particles called **atoms**. This second idea is now accepted as being correct and your cutting of the piece of iron would eventually come to a stop when you had left just one atom of iron. This is the basic building block of all iron objects. Atoms are so small that even under the most powerful of microscopes, we are still unable to see them.

Altogether, there are over 90 different atoms. Some of the objects you see around you will be made from just one type of atom. For example, objects made from copper contain just copper atoms and objects made from aluminium contain just aluminium atoms. However, most of the objects you can see have building blocks that consist of not one atom but several different atoms fixed together. For example, the building block for water is one atom of oxygen fixed to two atoms of hydrogen, and carbon dioxide has a building block that consists of one atom of carbon fixed to two atoms of oxygen. When several atoms are joined together like this they form a **molecule**.

Test Yourself

1   Explain the difference between the ideas about matter put forward by Aristotle and Democritus?

2   What are the basic building blocks for a) iron objects and b) copper objects?

3   What is the difference between an atom and a molecule?

4   Approximately how many different types of atoms are there?

Solids, liquids and gases

Although scientists now believe that everything around us is made up from these small particles we call atoms or molecules, there are still some questions to be answered about how these particles are arranged. Are they arranged differently in solids, liquids and gases? Are they stationary or are they moving? If we cannot see these particles, how are we ever going to answer these questions?

Good scientists are like detectives. Detectives normally do not see a crime take place but, by carefully examining the evidence, they can draw conclusions about what happened and perhaps 'who dunnit'. Scientists have tried to do something similar. They have studied all the available evidence and have formed a theory called the **kinetic theory**. This theory explains how the particles in the three different **states of matter** behave.

The evidence for a solid

1   One of the main properties of a solid is that it is SOLID. It has its own shape and can often support other objects. The stool you are sitting on and the desk you are leaning on are supporting your weight. They are able to do this because they are solids. Imagine what would happen if your stool or desk was made of a liquid or a gas! This behaviour suggests to scientists that the particles that make up a solid are somehow 'fastened together' so they cannot move apart when we lean on them.

2   Many solids, when they form from liquids, have their own particular shapes. We describe these solids as being **crystalline**. The regular shapes of these crystals suggest that the particles from which they are made have arranged themselves in an ordered fashion e.g. in straight lines.

Figure 1 ▲   These crystals of sodium chloride (table salt) are extremely regular in shape

3 Further evidence that the particles of a solid are arranged in straight lines appears when we try to cut crystals. This is called **cleaving**. If the cut is made between the lines of particles, the crystals cleave easily producing two flat surfaces. If, however, the cut is made in another direction, the crystal may shatter.

Before diamonds are put into a piece of jewellery, they must be cut into shape. The diamond cutter is extremely skilled at spotting the directions in which he can cut between the particles. If he gets it wrong, his mistake could be very expensive!

4 When we heat a solid, it usually expands – the particles spread out. When we cool a solid it usually contracts – the particles move closer together (see page 70).

5 Many solids can be stretched or compressed.

6 There are electrostatic charges within particles which can create attractive and repulsive forces.

7 The density of most solids is quite high compared with the densities of liquids and gases.

The model of a solid suggested by the above evidence

We believe that the particles of a solid are arranged in a pattern called a **lattice** and that there are electrostatic forces holding this structure together. The particles are able to move from side to side, i.e. vibrate, but they are unable to change their position within the lattice. They occupy fixed positions. As the temperature of a solid increases, the vibrations of its particles become more vigorous. If the temperature is decreased, the vibrations become less vigorous. At one particular temperature, called **zero kelvin** or **absolute zero** ($-273\,°C$), it is believed that all vibrations will stop.

We describe how vigorously particles are vibrating using the phrase **kinetic energy**. The more vigorously the particles of a solid vibrate, the greater its kinetic energy.

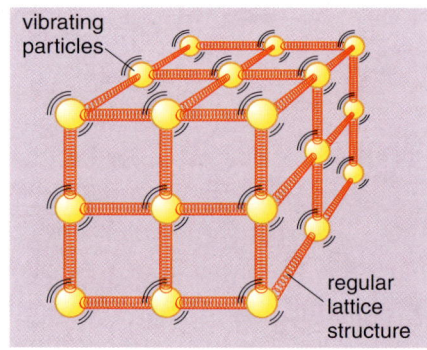

Figure 2 ▲ The model of a solid at room temperature

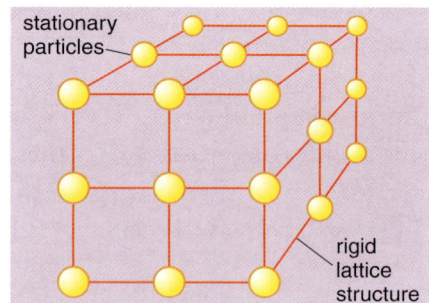

Figure 3 ▲ At zero kelvin (absolute zero) the particles of a solid stop vibrating

Test Yourself

5 What is a lattice structure?

6 Why are the particles in a solid unable to move around?

7 Why do most solids have a high density?

The evidence for a liquid

1 We know that one of the main properties of any liquid is its ability to **flow**. This suggests that the particles in a liquid do not have fixed positions. They can move about and therefore have more freedom than the particles in a solid.

2 Liquids take the shape of the container into which they are poured. They have no shape of their own. This suggests that there is no regular structure (lattice) within a liquid and therefore the attractive forces between particles are weaker than those in a solid.

3 It is possible with a little care to float a small steel needle or pin on the surface of water. If we look carefully at water, which is dripping slowly from a tap, we will see that it forms large droplets before it falls. Both of these observations suggest that there are still forces of attraction between the particles of a liquid.

4 If a small amount of fruit syrup is added to a beaker of water, we can see that it gradually spreads throughout the beaker even though it has not been stirred. This suggests that the particles that make up the water are not stationary but are slowly moving around. The movement of the water particles gradually spreads the syrup. This process of mixing is called **diffusion**.

5 The density of a liquid is usually less than that of a solid but more than that of a gas.

The model of a liquid suggested by the above evidence

It is not possible to draw an exact model for a liquid as we can for a solid, as the particles do not have fixed positions within a lattice type structure. The particles are vibrating and moving around. We believe that the distances between particles are a little larger in liquids than in solids and so the forces of attraction between the particles are weaker. As a result, the particles in a liquid are able to slide past each other i.e. they can flow.

liquids can flow

Figure 4 ▲   A liquid will take the shape of its container

Figure 5 ▲   A needle will float on the surface of water

Figure 6 ▲   When left alone, the syrup gradually diffuses throughout the water

Test Yourself

8 Why is it not possible to draw an exact model of a liquid?

9 Why are liquids able to flow?

10 What is diffusion in liquids?

The evidence for a gas

1 If we introduce a gas such as air into a container, it will fill all the space available no matter what its shape or size. This suggests that the particles of a gas are completely free from any forces trying to hold them together.

2 If we introduce a small amount of smoke into a smoke cell and observe the smoke particles through a microscope, we will see that they are not stationary. They move around inside the smoke cell in a very random (haphazard) manner. We believe that this motion, which we call **Brownian motion**, is being caused by the air particles inside the cell striking the smoke particles and pushing them around. If this is true, the air particles must be moving very fast in order to be able to move comparatively much larger objects such as the smoke particles. Imagine trying to move a large block of iron by shooting grains of rice at it. The grains would have to be moving very quickly.

3 Because the smoke particles in a Brownian smoke cell are being made to move in all directions, the air particles must also be moving in all directions.

4 It is very difficult to squash a solid or a liquid because the particles are close together. It is very easy to squash a gas. Also the density of a gas is much lower than that of a solid or a liquid. Both of these observations suggest that the particles in a gas are much further apart than they are in a solid or a liquid.

5 If a gas jar of air is placed on top of one of brown nitrogen dioxide, the two gases gradually mix i.e. they diffuse into each other. This diffusion takes place far more quickly than that in the syrup and water experiment (see page 63), suggesting that gas particles move around much more quickly than particles in a liquid.

Figure 7 ▲   A gas will completely fill its container

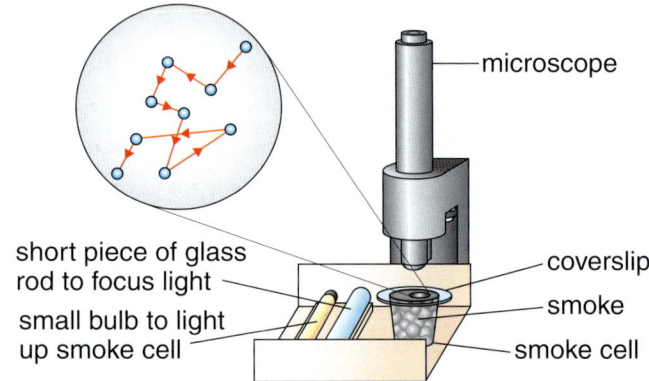

short piece of glass rod to focus light

small bulb to light up smoke cell

microscope

coverslip

smoke

smoke cell

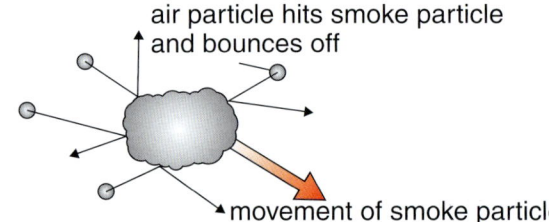

air particle hits smoke particle and bounces off

movement of smoke particle

Figure 8 ▲   Brownian motion

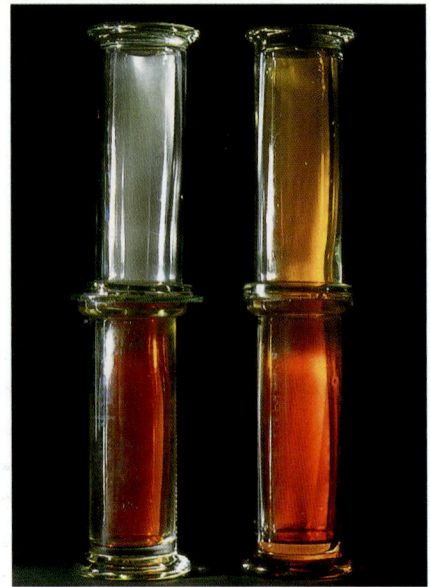

Figure 9 ▶   Diffusion in gases is much faster than diffusion in liquids

The model of a gas suggested by the above evidence

We believe that like liquids, gases have no fixed structure. The particles of a gas are widely spaced, are completely free from forces of attraction and are moving around at very high speeds (typically 500 m/s at room temperature) and in all directions.

Inside a container, such as a tyre, the gas particles are continually colliding with the sides. It is these collisions that create pressure inside the tyre. If more air is pumped into the tyre, there will be more particles and more collisions, so the pressure in the tyre will increase.

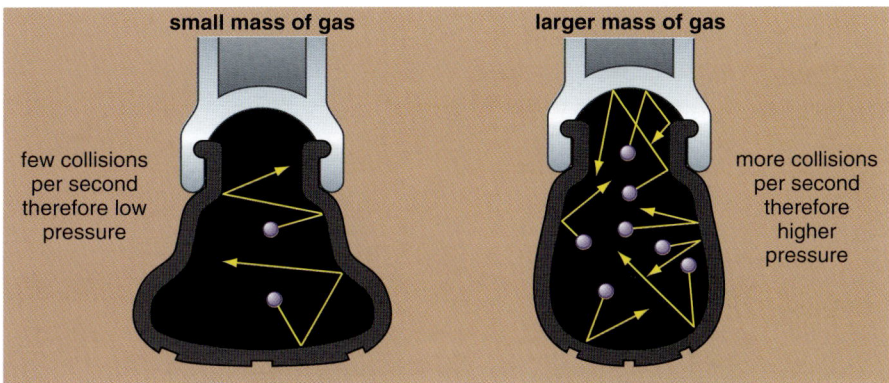

small mass of gas   larger mass of gas

few collisions per second therefore low pressure

more collisions per second therefore higher pressure

Figure 10 ▲   With more air in the tyre, the pressure is greater

Test Yourself

11 What two conclusions about the behaviour of gas particles can we draw from Brownian motion?

12 Why are gases easier to compress than solids and liquids?

13 Why is gaseous diffusion faster than diffusion in liquids?

Melting and boiling

Melting

If a solid is heated gently, its particles vibrate more and more vigorously. Eventually they vibrate so violently that the regular lattice structure breaks apart and the particles are able to move past each other. The temperature at which this happens is called the **melting point** of the solid. The solid has become a liquid. If we stop heating the liquid it will lose energy to the surroundings and the lattice structure will begin to reform. The liquid is **solidifying** or **freezing**.

Boiling

If a liquid is heated, its temperature will increase until the vibrations of its particles become so violent that they break away from neighbouring particles becoming completely free. The temperature at which this happens is called the **boiling point** of the liquid. The liquid has become a gas. If we stop heating the gas it will lose energy to its surroundings and forces of attraction will pull the particles together. The gas is **liquifying** or **condensing.**

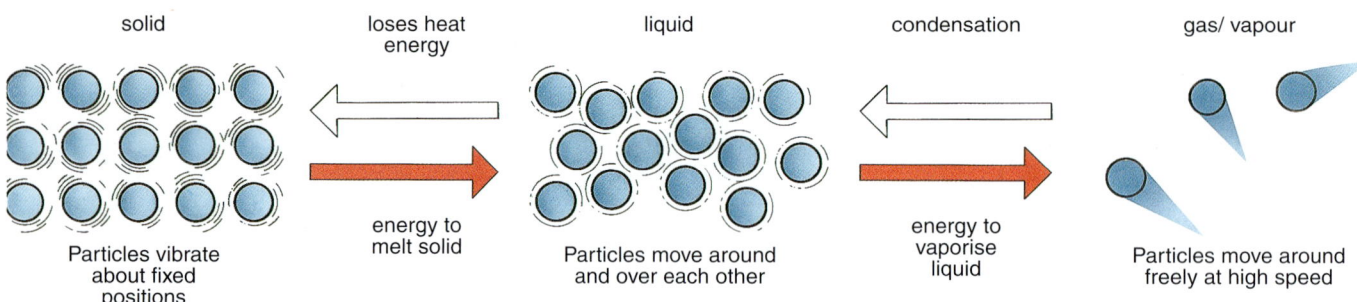

solid   loses heat energy   liquid   condensation   gas/ vapour

Particles vibrate about fixed positions

energy to melt solid

Particles move around and over each other

energy to vaporise liquid

Particles move around freely at high speed

Figure 11 ▲   Changes of state

Table 1 gives some examples of the melting points and boiling points of different substances.

Substance	Melting point/°C	Boiling point/°C
water	0	100
alcohol	−117	78
tungsten	3410	5500
iron	1539	2887
oxygen	−219	−183

Table 1 ▲   Comparing melting and boiling points of some common substances

Extension box

Figure 12 shows how the temperature of a substance which is initially a solid changes as it is heated until it becomes a gas.

Figure 13 shows how the temperature of a substance which is initially a gas changes as it cools and eventually becomes a solid.

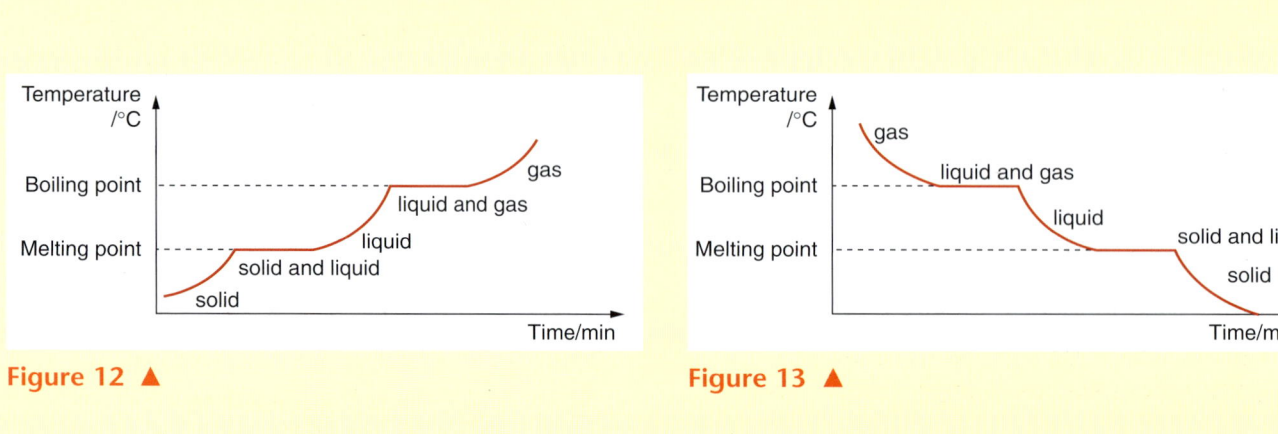

Figure 12 ▲

Figure 13 ▲

Summary

When you have finished studying this chapter, you should understand that:

✔ Solids, liquids and gases are the three states of matter. They have different physical properties such as shape, strength, density and ability to flow.

✔ These different properties can be explained by the arrangement and movement of their particles.

✔ The particles of a solid are close together, held in fixed positions by strong forces, and are vibrating.

✔ If the particles are arranged in a regular structure, such as a lattice, the solid may be crystalline.

✔ The particles in a liquid are a little further apart than in a solid, do not have fixed positions but are able to move around slowly. There are still forces of attraction between the particles.

✔ The particles in a gas are very far apart. The forces of attraction between them are negligible so they are able to move around at high speeds.

✔ Substances can change from one state to another, for example solids can melt to liquids and liquids can boil to gases.

End-of-Chapter Questions

1 Explain in your own words the following key terms you have met in this chapter:

atom	kinetic energy
molecule	flow
kinetic theory	diffusion
states of matter	Brownian motion
crystalline	melting point
cleaving	solidifying/ freezing
lattice	
zero Kelvin or absolute zero	boiling point
	liquifying/ condensing

2 a) What are the three states of matter?

b) Explain the differences in the movement of particles in these three states.

3 a) Draw a diagram of the model of a solid at room temperature.

b) Draw a model of a solid at zero Kelvin.

c) Explain in your own words the differences between the two models.

d) What types of forces are holding these models together?

e) What proof do we have that the particles in a solid are arranged in rows?

4 a) Why are most liquids incompressible?

b) Compare the freedom of movement of the particles in a liquid with that in solids and gases.

5 The diagram below shows gas particles inside a cylinder

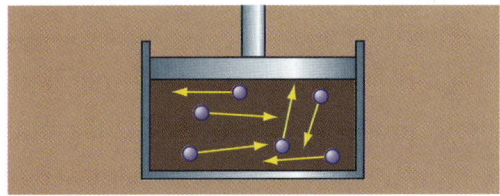

a) Explain how the gas particles create pressure inside the cylinder.

b) Explain what happens to this pressure if some of the particles are removed.

c) Explain what happens to the pressure if the piston is pushed downwards, reducing the volume the gas occupies.

7 Temperature and the movement of energy

This girl has her left hand in a bowl of hot water. It feels hot because energy is flowing *from* the water *into* her hand.

At the same time she has her right hand in a bowl of icy water. This water feels cold because energy is flowing *from* her hand *into* the water.

Energy is moving because there is a temperature difference. The **temperature** of an object or substance is a measure of its 'hotness'. The water in the bowl on the left is at a higher temperature than the girl's hand so heat flows from the water into her hand. The water in the bowl on the right is colder than her hand so heat flows in the opposite direction.

Test Yourself

1   What causes heat energy to move?

2   In which direction does heat energy move?

Measuring temperature

Human beings are not very good at judging temperatures. If the girl in the example above was to move her hands into a third bowl containing lukewarm water, her left hand would feel cold, whilst her right hand would feel warm. Our senses are comparing the temperatures of the bowls of water. They are not able to give an accurate value for the temperature. To measure the temperature of an object or substance accurately we use a **thermometer**.

There are several different types of thermometer shown below.

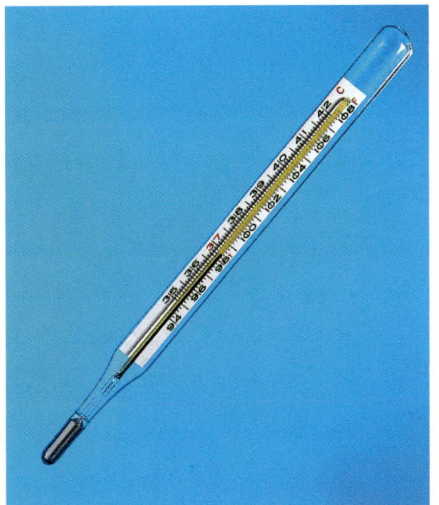

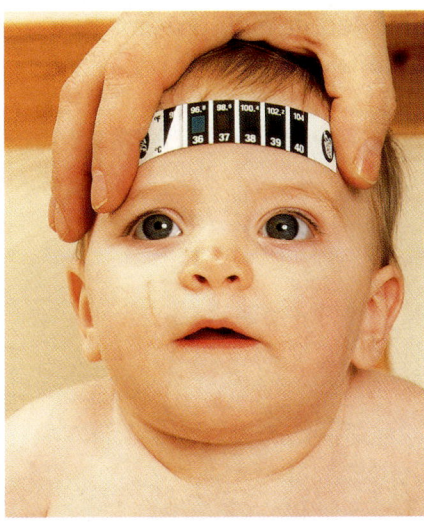

Figure 1 ▲ As the liquid mercury becomes warm, it expands and rises. The hotter it is, the further it rises. A scale at the side of the column shows the measured temperature

Figure 2 ▲ As the temperature of the plastic strip increases or decreases, its colour changes. The colour of the strip indicates its temperature

Figure 3 ▲ Temperature differences between wires inside a thermocouple create small voltages. The sizes of these voltages indicate the temperature

Examples of temperatures

We measure temperatures in **degrees Celsius** (°C).

Where	Temperature	Where	Temperature
absolute zero	−273 °C	boiling water	100 °C
coldest place on Earth	−89 °C	oven	200 °C
water freezes	0 °C	Bunsen flame	1100 °C
average room temperature in UK	20 °C	filament bulb	3500 °C
our body temperature	37 °C	surface of Sun	6000 °C
maximum temperature on the Earth	60 °C	centre of Sun	15 000 000 °C

Table 1 ▲ Comparing temperatures

Test Yourself

3 What piece of apparatus do we use to measure the temperature of an object?

4 Explain how you would measure the temperature of a beaker of water using a mercury in glass thermometer.

5 What is the approximate temperature in degrees Celsius of a) a fridge, b) a freezer and c) a burning match?

Temperature and heat (energy)

Although temperature and heat are related, it is important that you understand the difference between them.

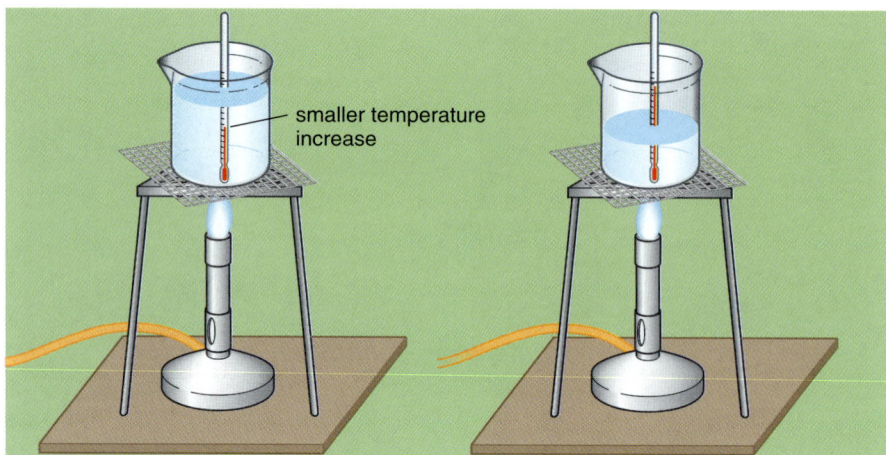

Figure 4 ◀

In Figure 4, gas is being burned by the Bunsen burner. Some of the energy released is transferred to the water causing its temperature to increase. If the experiment is repeated with a beaker containing twice as much water, the energy given to the water over the same length of time will be identical but the temperature increase will be smaller. It is clear from the above that heat is the energy given to the water and a temperature increase is the result of this transfer. We measure energy in **joules** (J) and temperature in **degrees Celsius** (°C).

Test Yourself

6   Explain in your own words the difference between heat and temperature.

7   In what units do we measure a) heat and b) temperature.

Expansion and contraction

In the summer these transmission lines seem to be quite slack. In the winter they appear to be very tight. It is the difference in temperature between the summer and the winter which causes this change. In the summer the cables become warm and **expand**. In the winter they cool and **contract**.

We can explain why the cables expand and contract as their temperatures change using the particle model from the previous chapter.

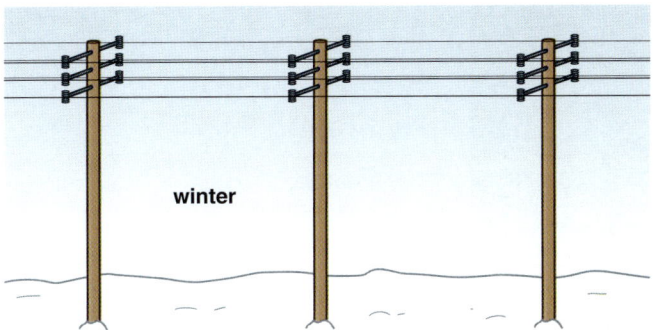

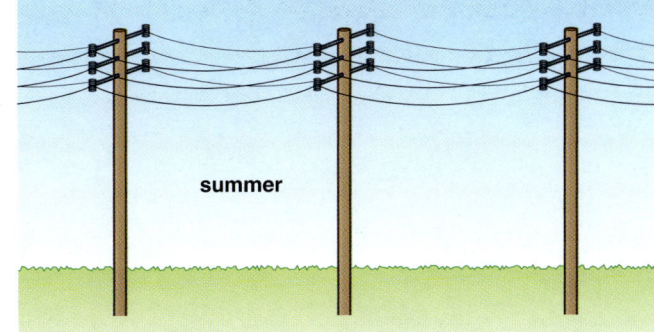

Figure 5 ▲ Comparing transmission lines in the summer and winter

When the cables are warmed they absorb energy. This causes the particles from which they are made to vibrate more vigorously. As a result of this 'extra' movement, the particles spread out a little and so the cables expand. In the winter the cables cool, the vibrations of their particles decrease and the cables contract.

We can demonstrate expansion and contraction due to changes in temperature in the laboratory using a ball and ring as shown in Figure 7.

At room temperature the ball just fits through the ring. When the ball is heated it can no longer pass through the ring, showing that it has expanded. When the ball is left to cool to room temperature, it is again able to pass through the ring, showing that it has contracted.

Forces of expansion and contraction

When objects expand and contract due to temperature changes, the forces involved are very large. This can sometimes be an advantage and sometimes a disadvantage.

When two metal plates are riveted together, the rivets must first be heated until they are extremely hot (white hot). The rivets are then inserted into holes drilled in each plate and then hit with a special hammer to create a second head. As they cool, the rivets contract pulling the two plates together. The forces involved are so strong that the joint between the plates is watertight. This is why the metal plates used to build ships are riveted together.

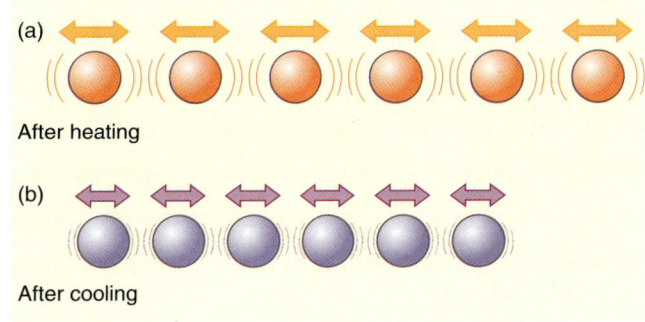

(a) After heating

(b) After cooling

Figure 6 ▲

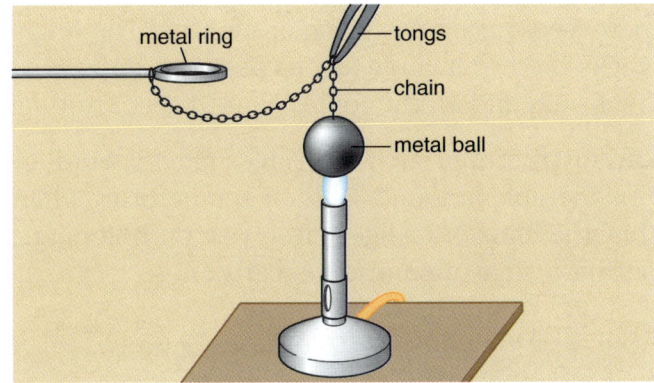

metal ring — tongs — chain — metal ball

Figure 7 ▲ The ball and ring experiment

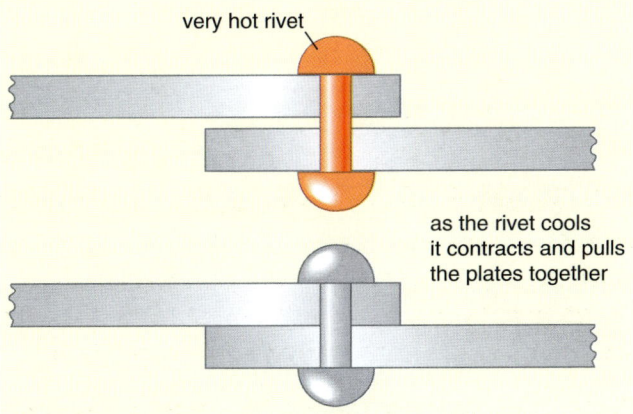

very hot rivet

as the rivet cools it contracts and pulls the plates together

Figure 8 ▲ As the rivet cools it contracts, pulling the two plates together

71

Figure 9 ▲ These railway tracks buckled in the hot summer

The railway lines in Figure 9 tried to expand as their temperature increased during the summer months. The engineers who laid the tracks did not include any **expansion gaps**. As a result, as the tracks expanded, the forces caused them to buckle

Large structures such as bridges may expand and contract by considerable amounts as their temperature changes. To allow for this, the ends of bridges often rest on rollers and small gaps permit a small amount of expansion.

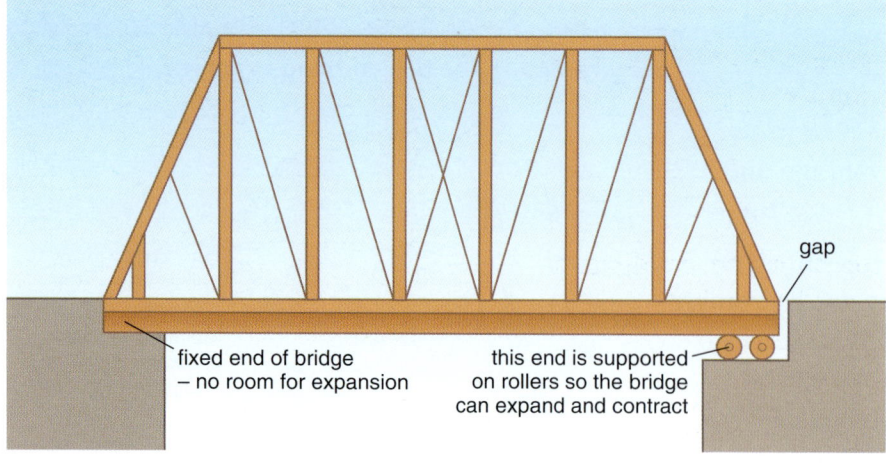

Figure 10 ▲ The expansion rollers on this bridge prevent buckling

The Firth of Forth railway bridge is approximately 1 m longer in the summer than it is in the winter.

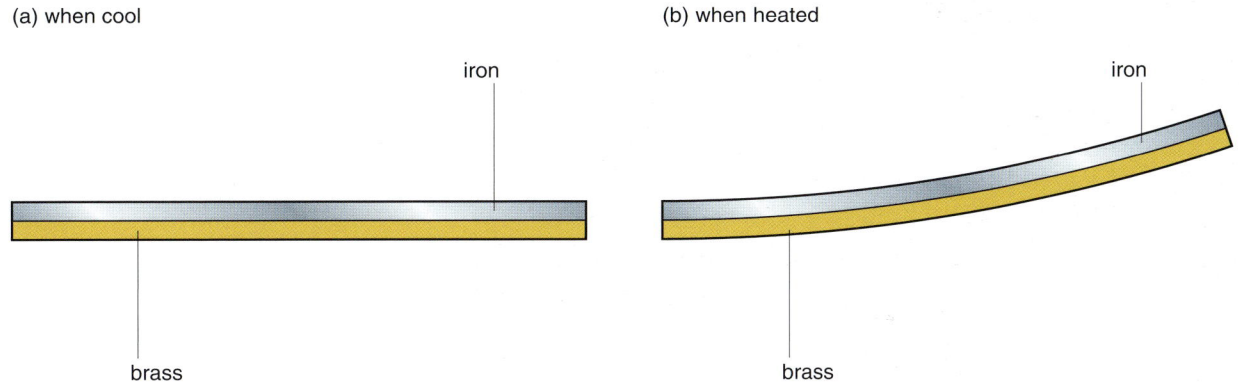

(a) when cool   (b) when heated

iron

iron

brass

brass

Figure 11 ▲   The bimetallic strip bends when heated and becomes straight again when cool

The bimetallic strip

Different materials can expand by different amounts when heated. This idea is used in the **bimetallic strip**. A bimetallic strip consists of two different metals stuck or riveted together. One of the metals expands a lot when heated. The other, when heated through the same temperature difference, expands very little. As a result when heated, the bimetallic strip bends. The metal which expands the most is on the outside of the bend. When the strip cools it becomes straight again.

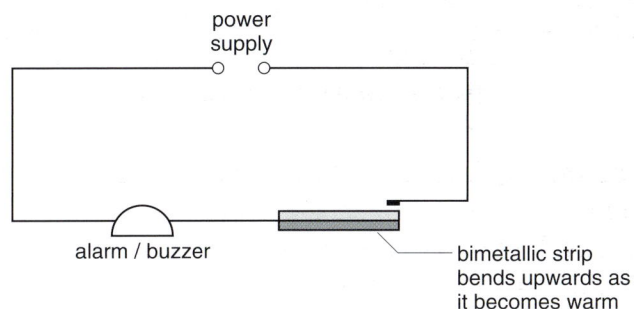

power supply

alarm / buzzer

bimetallic strip bends upwards as it becomes warm

Figure 12 ▲   This fire alarm circuit makes use of a bimetallic strip

The bimetallic strip is often used in the circuits of thermostats and fire alarms (see Figure 12).

If there is a fire and the strip becomes hot, it bends, completes the circuit and the alarm bell rings.

Heat – energy on the move

There are several ways in which heat can move: conduction, convection and radiation.

1 Conduction

When you warm up food in a saucepan over a heater, energy needs to travels from the heater through the base of the pan to the food. The heat energy moves by **conduction**.

The particles that form the outside of the base of the pan gain energy from the heater and are made to vibrate more vigorously.

Neighbouring particles within the pan are then jostled by these excited particles causing them to vibrate more vigorously. Eventually all the particles in the base of the pan will be vibrating vigorously, i.e. heat has been conducted through the base.

Because we want heat to travel easily and quickly through the pan, it is made from a material which is a good **conductor** of heat such as steel or copper. All metals are good conductors of heat.

Test Yourself

11 Give one example of a material which is a) a good conductor and b) an insulator.

12 Explain in your own words how heat is conducted through a metal rod which is being heated at one end.

13 Explain why materials such as fibre glass are really good insulators.

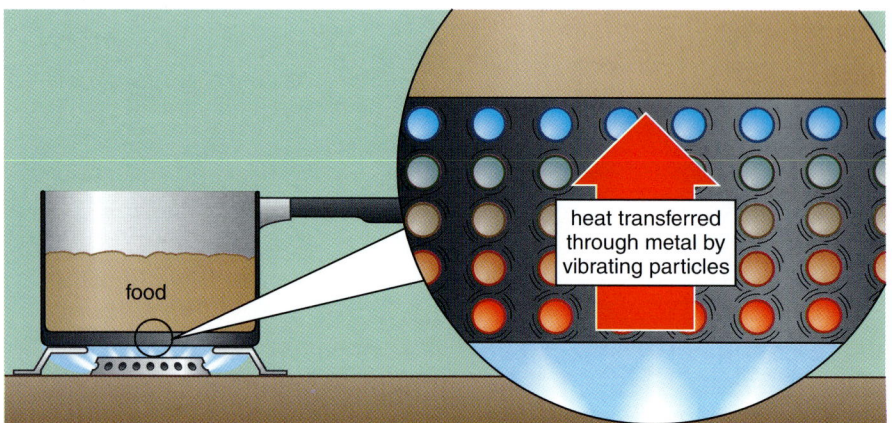

Figure 13 ▲ A particle model of conduction

heat transferred through metal by vibrating particles

food

To be able to pick up a pan without burning your hand, heat must not be allowed to travel into the handle. The handle must therefore be made from a material that does not allow heat to travel through it easily. A material which prevents or reduces conduction of heat is called an **insulator**. Examples of insulators include plastics, rubber and wood.

Most liquids are poor conductors of heat. The experiment in Figure 14 demonstrates that water is a very poor conductor of heat. The ice at the bottom of the boiling tube melts very slowly even though there is boiling water just a few centimetres above it. The water between the top and the bottom is a very poor conductor and so little heat is able to move between the two.

Gases are extremely poor conductors of heat. In fact they are excellent insulators. They are often used in situations where we wish to reduce heat loss.

- Double glazing traps a layer of air between the two panes of glass to reduce heat loss through the window.

- Woolly hats and jumpers trap air which reduces heat loss from the body, keeping us warm.

- Fibre glass is made of lots of trapped pockets of air which reduce heat loss through ceilings and walls.

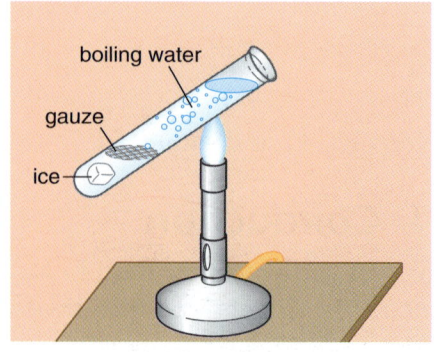

boiling water

gauze

ice

Figure 14 ▲ This simple experiment shows that water is a poor conductor of heat

2 Convection

Heat is transferred from a heater to all parts of a room by **convection**.

The air near the heater becomes warm and expands. It is now less dense than the air around it and so rises. Cooler, more dense air moves in to take the place of the warmer air. Away from the heater the air cools, becomes more dense and sinks. This circular movement of air is called a **convection current**.

The transfer of heat by convection can take place in liquids and gases but not in solids.

As the water immediately above the flame becomes warm, it expands and rises taking energy around the beaker. A small amount of colouring added to the water makes it easier to see these movements.

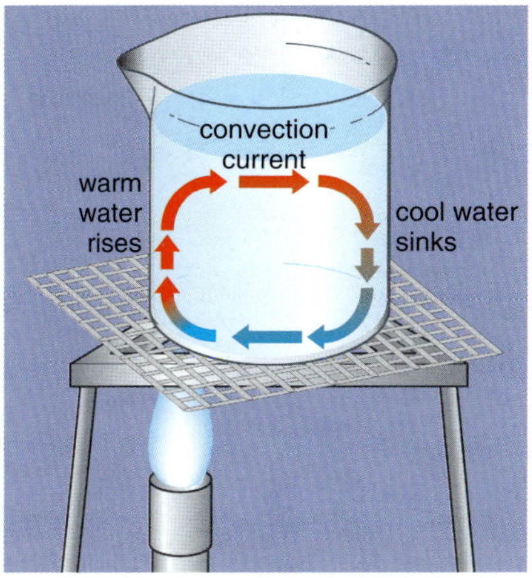

Figure 15 ▲   The heat from the Bunsen burner has set up a convection current in this beaker of water

Test Yourself

14 What is a convection current?

15 Where in a beaker of water is the water warmest? Explain your answer.

3 Radiation

The heat travelling to us from the Sun moves by radiation. It cannot be transferred by conduction and convection as there are no particles between the Sun and the Earth's atmosphere – it is a vacuum. Heat can only be transferred across a vacuum by radiation.

When this radiation strikes an object it may be **absorbed** or **reflected**. If an object absorbs the radiation, it becomes warmer.

Figure 16 ◄   Heat energy travels out from this fire to the boy as waves. It travels by **radiation**. If someone were to stand between the boy and the fire, the radiation would be blocked off and the boy would receive no energy

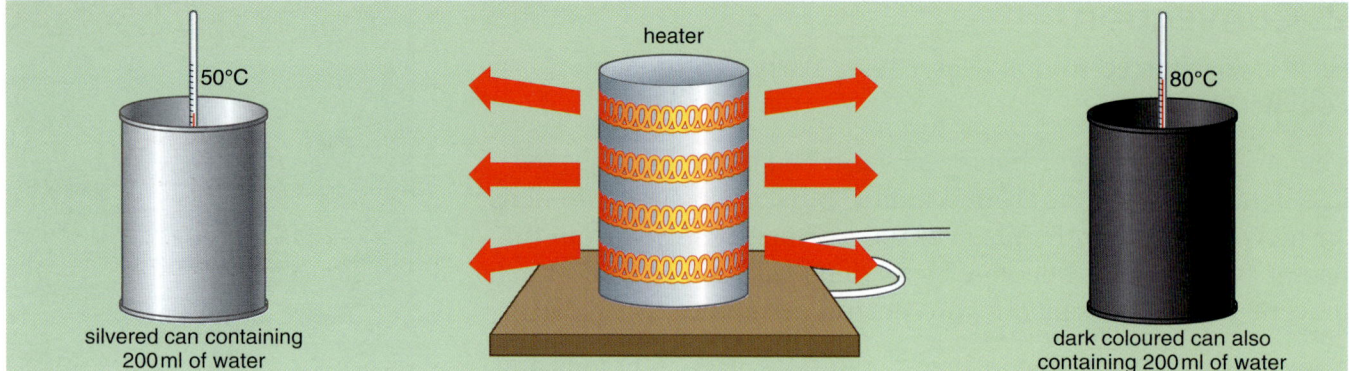

Figure 17 ▲ Objects that have dark rough surfaces absorb most of the radiation that strikes them and become warmer. Objects that are light coloured and smooth reflect most of the radiation and remain cool

Objects that have dark rough surfaces absorb most of the radiation that strikes them and become warmer. Objects that are light coloured and smooth reflect most of the radiation and remain cool.

Houses in hot countries are often painted white so that they reflect most of the radiation and so remain cool.

Heat loss from the home

Test Yourself

16 Which is the only way that heat can travel across a vacuum?

17 Name two things that may happen to radiation when it strikes an object.

18 Why are houses in hot countries often painted white?

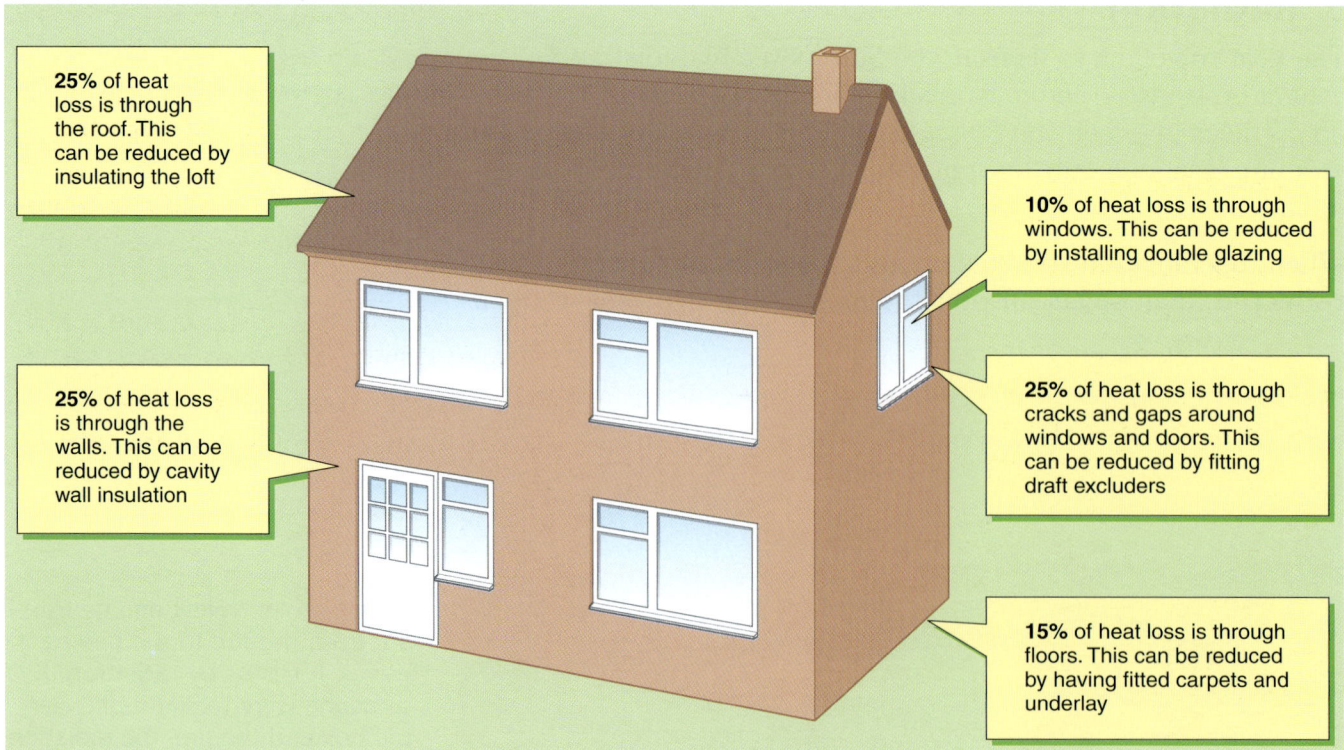

25% of heat loss is through the roof. This can be reduced by insulating the loft

10% of heat loss is through windows. This can be reduced by installing double glazing

25% of heat loss is through the walls. This can be reduced by cavity wall insulation

25% of heat loss is through cracks and gaps around windows and doors. This can be reduced by fitting draft excluders

15% of heat loss is through floors. This can be reduced by having fitted carpets and underlay

Figure 18 ▲ Different ways in which heat is lost from an uninsulated house

Summary

When you have finished studying this chapter, you should understand that:

✔ Temperature is a measure of the 'hotness' of an object or substance.

✔ To measure temperatures accurately we use thermometers.

✔ We measure temperatures in degrees Celsius (°C) and energy in joules (J).

✔ Objects/substances expand when heated and contract when cooled.

✔ Temperature difference causes heat energy to move.

✔ Heat energy can be transferred by conduction, convection or radiation.

✔ Objects that allow heat to pass through easily are called conductors.

✔ Objects that do not allow heat to pass through them easily are called insulators.

✔ Heat loss from the home can be reduced and energy conserved by using insulating techniques such as double glazing, loft insulation, cavity walls, carpets and underlay, and draft excluders.

End-of-Chapter Questions

1   Explain in your own words the following key terms you have met in this chapter:

temperature	conduction
thermometer	conductor
joules	insulator
degrees Celsius	convection
expand	convection current
contract	radiation
expansion gap	absorption
bimetallic strip	reflection

2   Explain what energy transfers take place when a piece of fresh meat is a) placed in a freezer and b) placed in an oven.

3   Explain why a stuck metal cap on a sauce bottle can sometimes be released by running hot water over it.

4   Explain why on a hot sunny summer day, a black car may be much hotter to touch than a white car parked next to it.

5   Explain why the heater in an oven is placed at the bottom but the freezing compartment in a fridge is placed at the top. Draw diagrams of each to aid your explanation.

6   Explain why heat can be transferred by convection through liquids and gases but not through solids.

7   Find out how the construction of a vacuum flask helps to prevent heat escaping from inside.

8   Find out a) what a thermostat is and b) how a bimetallic strip is used inside a thermostat.

9   Find out why onshore sea breezes occur during the daytime but offshore sea breezes occur at night.

Rays of light and reflection

What do all the following have in common: closing your eyes, putting your hands over your eyes, wearing a blind fold and turning off the lights?

It should not take you too long to realise that they are all different ways of stopping you from seeing things. In order to see we need light and this light must enter our eyes.

We see some objects like the Sun, stars and fires because they emit light. This light then travels into our eyes. Objects like these are called **luminous objects**. Other objects such as dogs, cats, tables and chairs do not emit their own light. They are **non-luminous objects**. We see non-luminous objects because of the light they reflect.

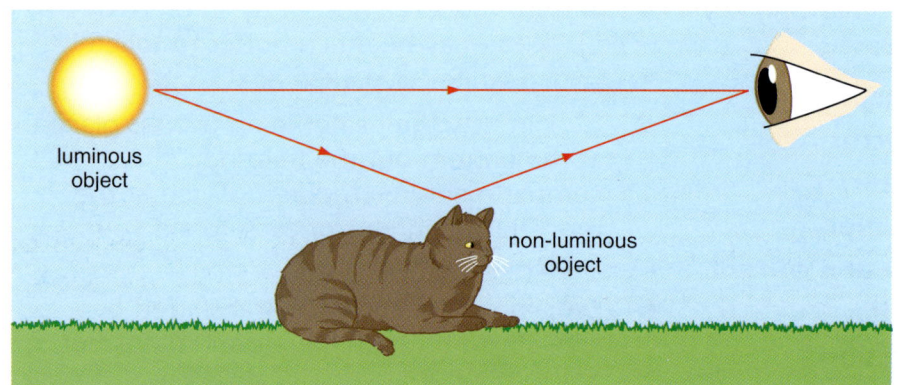

luminous object

non-luminous object

Figure 1 ◄ Some objects are luminous, others are non-luminous

Test Yourself

1   Name one object in the night sky which often provides us with light but is a non-luminous object. Can you explain how it does this?

2   Name one luminous object which is not hot. (*Hint*: It's alive)

How does light travel?

Walking through a thick forest on a bright sunny day might provide us with some clues as to how light travels. Beams of light which can be seen to be *travelling in straight lines* stream through the trees and their leaves.

Figure 2 ◀ Light travels in straight lines

Observing the patterns of light in a forest gives us just one clue that light travels in straight lines. There are several other experiments and observations that lead us to the same conclusion.

If we take a short length of rubber tubing and hold it so it is in a straight line, we should be able to see objects through the tubing. If, however, the tubing is curved or bent so that it is not straight, it is impossible to see objects at the far end of the tubing.

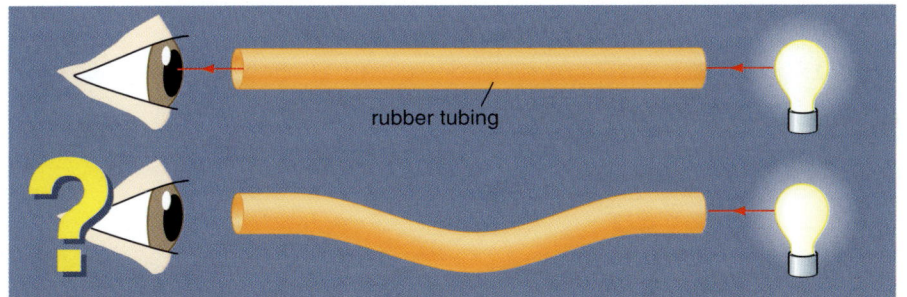

rubber tubing

Figure 3 ◀ How to demonstrate that light travels in straight lines

Test Yourself

3 Where on a misty night might you see further evidence that light travels in straight lines?

4 Even travelling in straight lines, it takes 8 minutes for light from the Sun to reach the Earth. If the distance from the Sun to Earth is 150 million kilometres, calculate in metres per second the speed at which light travels.

5 Explain in your own words why it is impossible to see around corners.

6 If you were given the following apparatus, how would you prove that light travels in straight lines?

a light source, three screens with a small hole in the same place on each screen and a length of cotton.

Drawing ray diagrams

Because light travels in straight lines we can draw **ray diagrams** to show and explain what happens to light when it strikes various objects. A ray is drawn as a straight line with an arrowhead showing the direction in which the ray is travelling. If the direction of the ray changes, it is usual to draw one arrowhead before the change in direction and one after the change.

Transparent, translucent and opaque

If we look through a pane of plain glass, we can see objects that are on the other side. See-through objects like this which allow light to pass through them are called **transparent** objects. If we replace the glass with a sheet of tracing paper and place a light source behind it, some of the light will pass through the paper but we would be unable to see through it. This happens because most of the light is scattered in new directions as it passes through the tracing paper. Materials which allow light to pass through them and yet do not allow us to see objects on the other side are described as being **translucent**. Most objects do not allow light to pass through them. They are described as being **opaque**.

Figure 4 ▲ Light can travel through transparent objects but not through opaque objects

Test Yourself

7 Give two examples of materials that are a) transparent, b) translucent and c) opaque.

8 Describe one domestic situation where you might use a material that is translucent.

Shadows

If we place an opaque object between a light source and a screen, we create an area of darkness where the light cannot reach. This area where there is a lack of light is called a **shadow**. On most occasions it is clear to see that a shadow has the same shape as the opaque object creating it. This again suggests that light travels in straight lines.

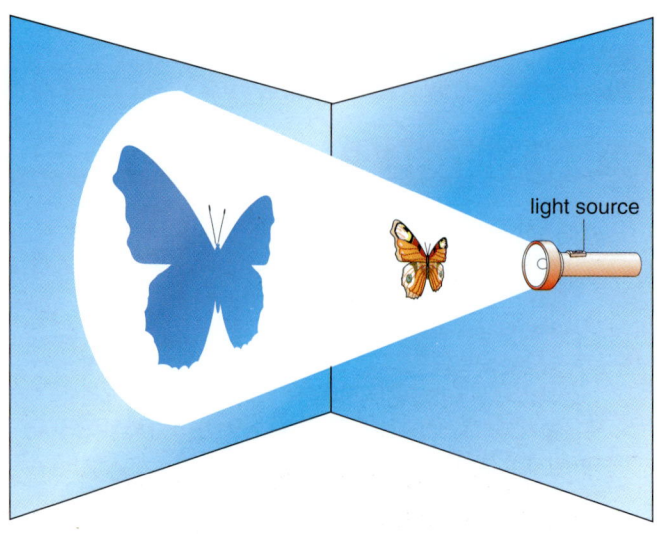

Figure 5 ▲ How an opaque object creates a shadow

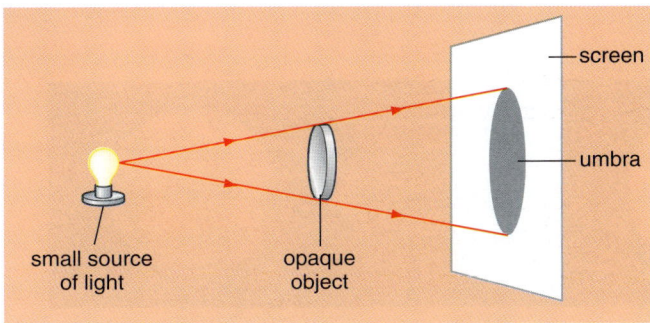

Figure 6 ▲   Creating a shadow

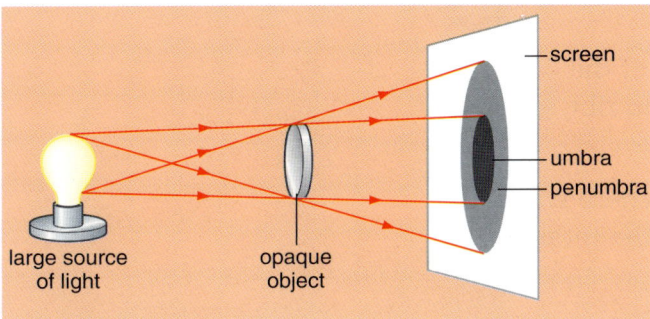

Figure 7 ▲

The ray diagram in Figure 6 shows how a shadow is created. This dark shadow is called an **umbra**.

If we replace the small light source with one which is a little larger (see Figure 7), we see two types of shadow on the screen. In the centre of the shape where no light is reaching the screen there is the dark umbra shadow. Around the edges of the shape there is a lighter shadow called a **penumbra**. The penumbra is created because only part of the light from the source is being blocked off.

Test Yourself

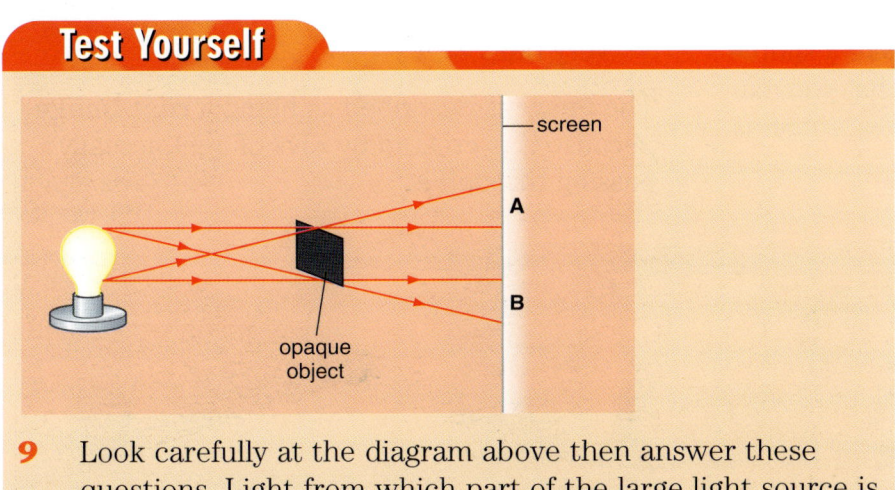

9   Look carefully at the diagram above then answer these questions. Light from which part of the large light source is able to reach the screen at a) A and b) B?

Extension box

One of the most spectacular examples of an opaque object creating umbra and penumbra shadows occurs when the Moon passes between the Sun and the Earth. This is known as a solar eclipse.

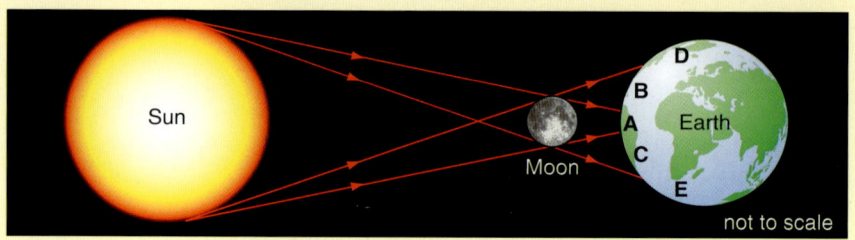

Figure 8 ▲ A solar eclipse

Test Yourself

10 If you were an observer of the eclipse in each of the regions marked A, B, C, D and E, draw and describe what you would see.

11 Find out when the next solar eclipse is due. Where will it be? Why will we in Britain not be able to see the total eclipse?

12 What is a lunar eclipse? Draw a ray diagram to show how it is different from a solar eclipse.

Ideas and Evidence

The pinhole camera

One of the earliest types of camera was the pinhole camera. As its name suggests, one side of the camera had a small pinhole in it to let in the light. On the far side was a photographic plate on which the image was created.

Figure 9 shows how you might make a simple version of the pinhole camera so that you can see how it worked and discover the problems of taking a photograph with it.

Ray diagrams show that the image created in the camera is

★ upside down
★ smaller than the object whose picture is being taken
★ very dull.

The image is described as being a **real image** because it is created by rays of light actually passing through it.

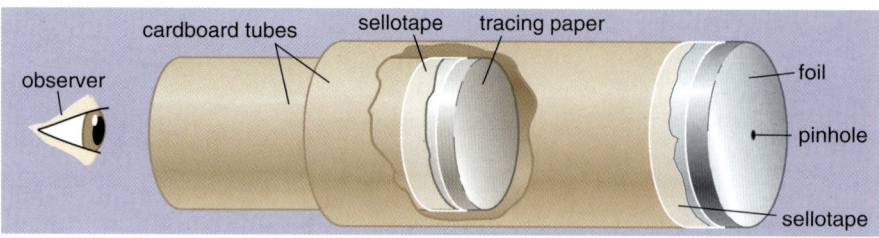

Figure 9 ▲ Making a simple pinhole camera

The pinhole camera continued

As the pinhole is small, very little light enters the camera. As a result it takes some time for enough light to enter the camera in order to produce a good image on the photographic plate. This is the reason why very early photographs always show posed, still subjects. If there is any movement whilst the plate is being exposed, the photograph is blurred.

Test Yourself

13 What do we call an image often created in hot countries which does not really exist? Find out how these images are created.

14 What is the exposure time for a camera? What determines how long an exposure should be?

To resolve this problem, more light needs to enter the camera and so the hole needs to be bigger. Unfortunately, this creates another problem. As the hole in a pinhole camera is made bigger, the image it creates becomes more blurred.

The problem was eventually solved by using **lenses**. These are specially shaped pieces of glass or plastic which are used to change the directions of rays of light in a controlled and predictable way. Even with large holes at the front of a camera, lenses are able to produce sharp images.

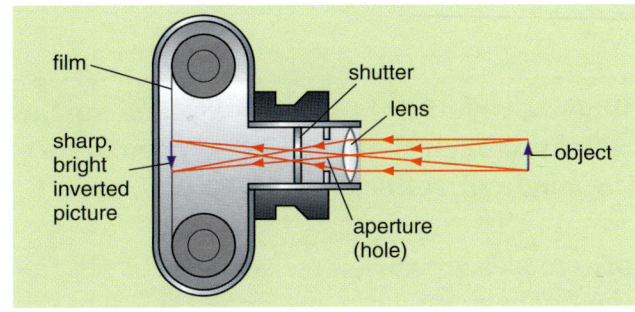

Figure 11 ▲   A modern camera has a lens to focus the light rays and produce a sharp image

Test Yourself

15 Describe the main features of the image created by a pinhole camera.

16 Why could a pinhole camera never be used to take action shots?

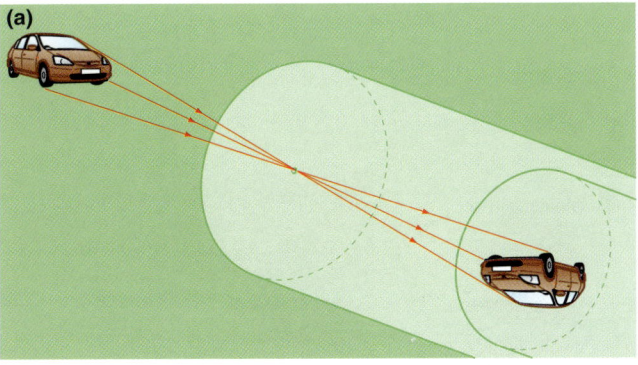
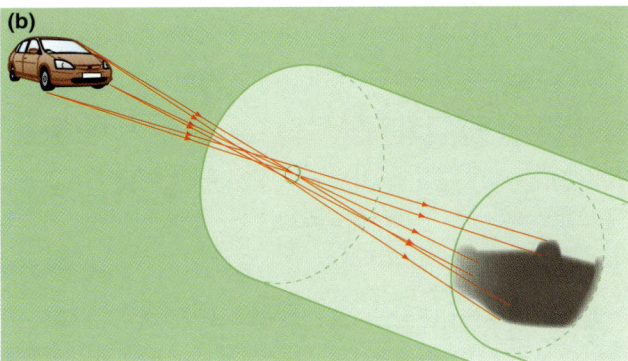

Figure 10 ▲   As the hole in a pinhole camera is widened, the image becomes blurred

Reflection from a plane mirror

If the old western films are to be believed, Red Indians did not just communicate with each other by sending smoke signals, they also used to send light signals. Not through optical fibres buried under the ground as we might do, but through the air using shiny pieces of metal or mirrors to reflect the Sun's rays. The law which describes these reflections states that

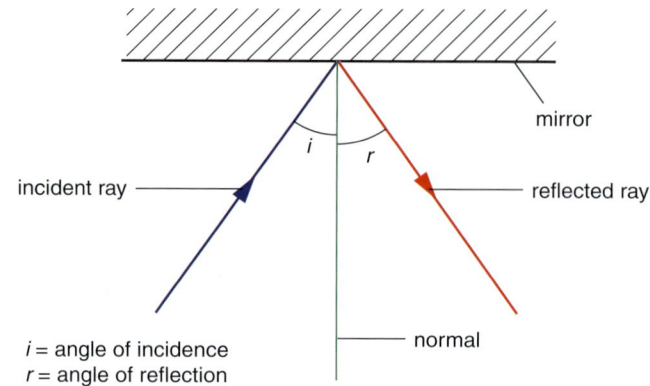

i = angle of incidence
r = angle of reflection

Figure 12 ▲ The angle of incidence *always* equals the angle of reflection

> **When a ray of light strikes a plane surface it is reflected such that the angle of incidence is equal to the angle of reflection.**

Note: Angles are always measured between the ray and the normal. The normal is a line drawn at 90° to the mirror. This avoids problems, for example if the mirror has a curved surface.

Test Yourself

17 Draw an accurate ray diagram showing what happens to a ray of light when it strikes a plane mirror at an angle of incidence of a) 25° and b) 75°.

The periscope

A simple periscope makes use of reflections from plane mirrors so that we can see over high objects and around corners.

Test Yourself

18 Describe with diagrams two situations where a periscope might be used.

Regular and diffuse reflection

Now we understand how rays of light are reflected from plane surfaces, we can perhaps explain why some flat surfaces appear shiny whilst others look matt or dull.

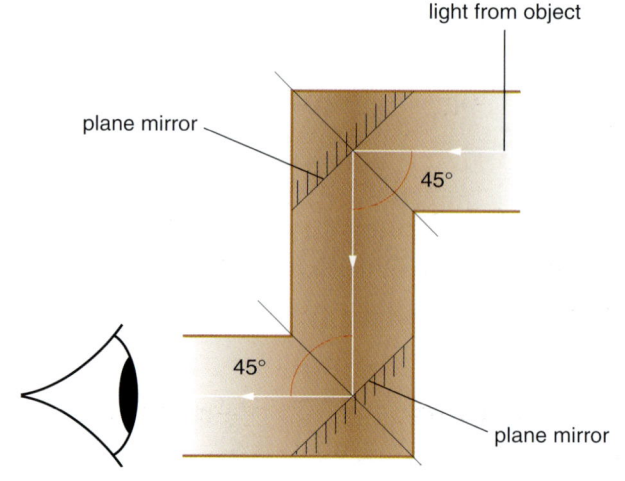

Figure 13 ▲ A simple periscope

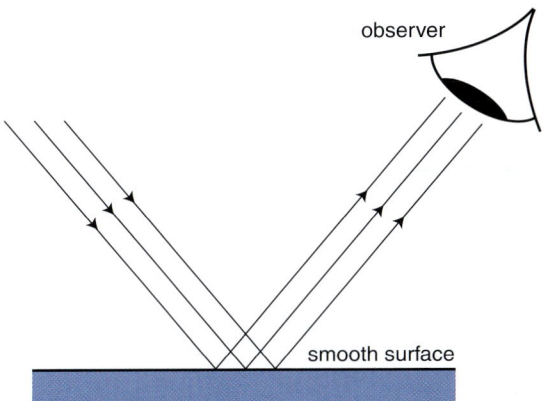 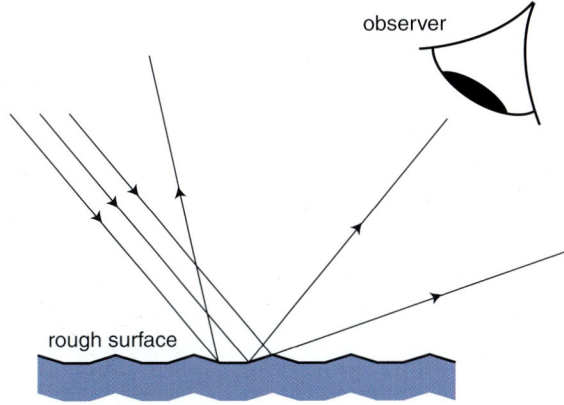

Figure 14 ▲   The difference between **regular reflection** and **diffuse reflection**

If a beam of parallel rays strikes a really flat smooth surface, all the rays will be reflected in the same direction. As a result many of the reflected rays will enter the eyes of an observer. To the observer the surface will therefore appear shiny.

If a beam of light strikes a surface which is not truly smooth, the rays will be reflected in different directions. As a consequence only a few of the rays will enter the eyes of the observer. The surface will therefore have a dull or matt appearance.

The image created in a plane mirror

When we look into a plane mirror we see images of ourselves and the objects around us, but how are these images created.

Rays of light from an object strike the mirror. Some are reflected into the eyes of the observer. Our brains are used to the idea that light travels in straight lines and so see the rays as having come from point I. This is where we see an image of the object. Images like this which have no rays of light actually passing through them are called **virtual images**.

The image created by a plane mirror is

- the same size as the object

- the same distance behind the mirror as the object is in front

- upright

- **laterally inverted** i.e. the object's left side is the image's right side

- virtual.

Test Yourself

19 Think carefully about the appearance of your leather shoes before and after you clean them. Can you now explain the difference in their appearance?

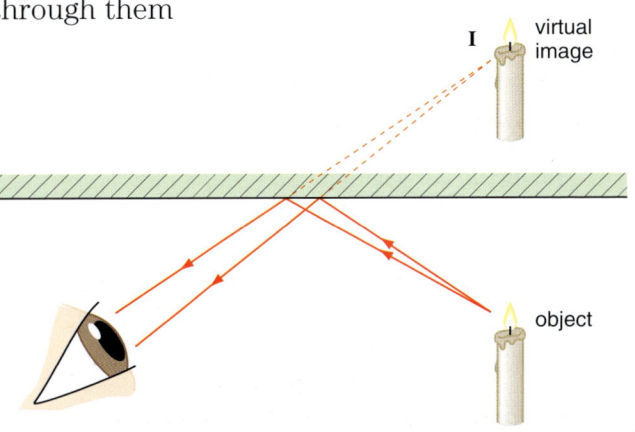

Figure 15 ▲   The image created in a plane mirror

Test Yourself

20 Can you think of a simple experiment to test an image to see if it is real or virtual. (*Hint*: You will need just a single piece of plain paper.)

21 Why has some of the writing on this ambulance been laterally inverted?

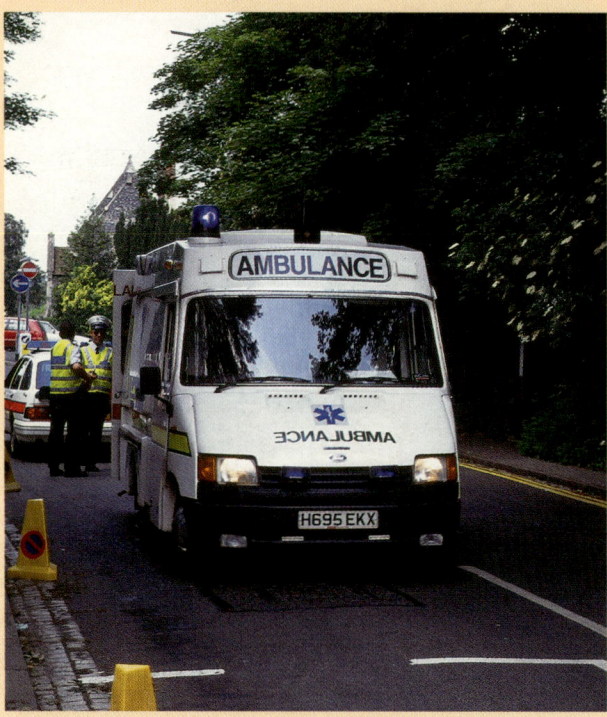

Summary

When you have finished studying this chapter, you should understand that:

✔ Light travels in straight lines.

✔ Light travels very quickly.

✔ We see luminous objects because of the light they emit.

✔ We see non-luminous objects because of the light they reflect.

✔ We can draw ray diagrams to show the movement of light.

✔ Transparent and translucent materials allow light to pass through them.

✔ Opaque materials do not allow light to pass through them. If an opaque object is placed between a light source and a screen, a shadow is created.

✔ Light is reflected from a plane surface at the same angle as it strikes the surface. In other words, the angle of incidence is equal to the angle of reflection.

✔ Rays of light actually pass through a real image.

✔ Rays of light only appear to pass through a virtual image.

End-of-Chapter Questions

1 Explain in your own words the following key terms you have met in this chapter:

luminous	penumbra
non-luminous	real image
ray diagram	lens
transparent	regular reflection
translucent	diffuse reflection
opaque	virtual image
shadow	laterally inverted
umbra	

2 A ray of light strikes a plane mirror at an angle of 60° to the mirror's surface.

a) What is the angle of incidence for this ray?

b) What is the angle of reflection?

c) What is the total angle through which the ray has been turned?

d) What would the angle of incidence be for a ray which is turned through 90°? Name an optical instrument which turns a ray of light through 90° twice.

3 A girl lies in front of a plane mirror. Her feet are 2 m from the mirror. Her head is 3.5 m from the mirror.

a) How far behind the mirror is the image of the girl's feet?

b) What is the total distance between the girl's head and the image of the girl's head created by the mirror?

4 It is possible to buy paints with a gloss finish or paints with a matt finish. Explain with diagrams the difference between these paints.

5 Write out a short poem or limerick in mirror writing. When you have finished, you should be able to place a plane mirror at the top of your poem and read all the words whilst looking in the mirror.

9

Refraction of light

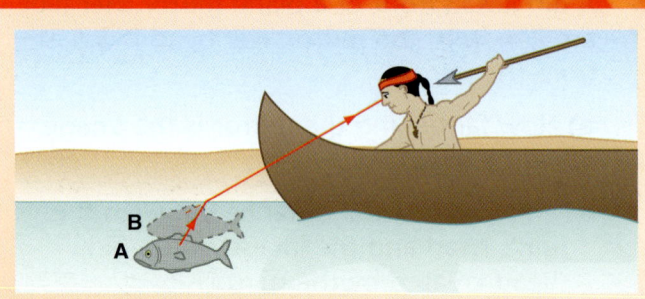

S pearing this fish is not quite as easy as it first seems. Rays of light from the fish change direction as they cross the water/air boundary. The fisherman therefore sees the fish at B rather than at A.

An experienced hunter will allow for this optical deception . . . an inexperienced one may not and so will miss the fish.

This change in direction of a ray of light as it travels from one **medium** to another is called **refraction**. (A medium (plural: media) is a transparent material through which a ray of light is travelling.)

Refraction

Refraction happens because light travels at different speeds in different media. In a vacuum (and in air), light travels at a speed of 300 million metres per second. It takes just 8 minutes for a ray of light to travel from the Sun to the Earth. In other media, such as water or glass, light travels more slowly.

When a ray of light travels into a glass block, it slows down and is refracted towards the **normal**. When a ray of light leaves a glass block and re-enters the air, it bends away from the normal.

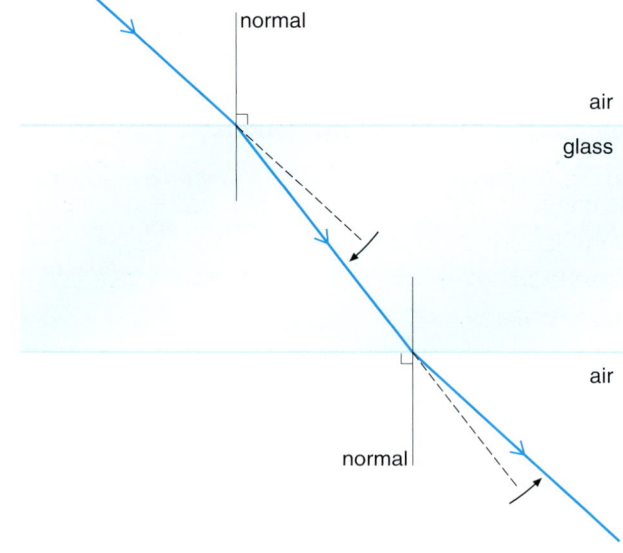

Figure 1 ▲   Refraction of light in a glass block

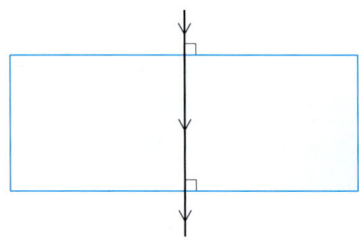

Figure 2 ◄   If a ray of light crosses the boundary at 90°, the ray is undeviated, i.e. it is not refracted

Figure 3 ▲   The pencil seems to bend at the air/water boundary

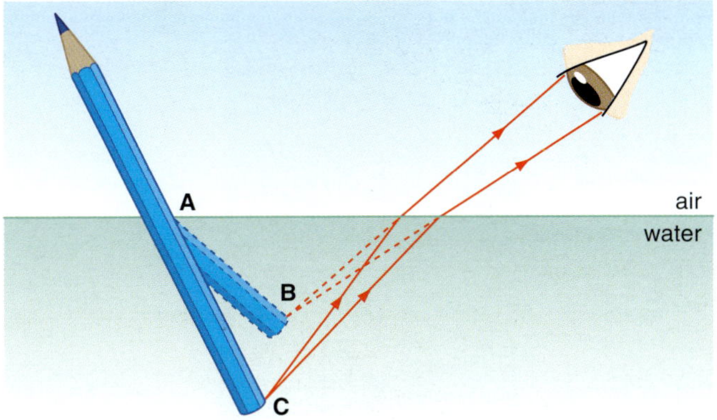

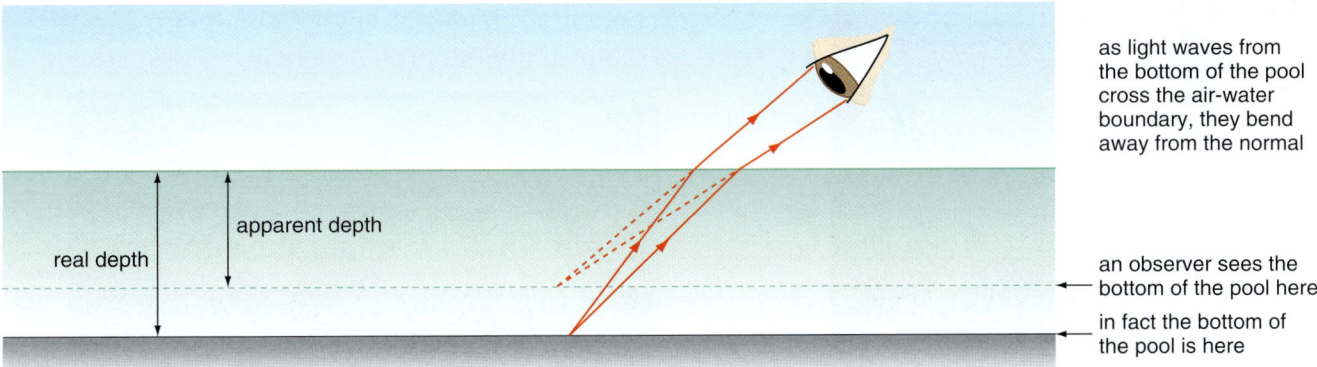

Figure 4 ▲   Why the pencil appears to be bent

These changes in direction due to refraction can cause some strange optical illusions. Figure 4 shows a pencil which appears to be bent.

Rays of light from the part of the pencil below the water bend as they cross the water/air boundary. The eye therefore sees this part of the pencil as being between A and B rather than between A and C.

Refraction also causes streams and ponds to look shallower than they really are.

When a ray of light passes through a sheet of glass, e.g. a window, it is refracted twice, once as it enters the glass and again as it emerges. However, because the pane of glass is quite thin, the effect is hardly noticeable. If, however we observe an object through very thick glass the effect is very noticeable (see Figure 6).

as light waves from the bottom of the pool cross the air-water boundary, they bend away from the normal

an observer sees the bottom of the pool here

in fact the bottom of the pool is here

Figure 5 ▲   The refraction of light at the air/water boundary makes this pond look shallower than it really is

Figure 6 ▲ This man is distorted when viewed through thick glass. He is standing behind the glass which is over a metre thick, but he looks much nearer

Colour

Ideas and Evidence

Sir Isaac Newton

In 1665 a young scientist called Isaac Newton was attempting to make a refracting telescope. (A *refracting* telescope uses lenses to produce an image. A *reflecting* telescope uses mirrors to produce its images.) But no matter how carefully he worked, the lenses he ground always produced coloured fringes around the images they created. Newton decided that he would need to study how these coloured rings were being produced before he could successfully produce the lenses he needed for his refracting telescope.

He carried out a simple experiment in a darkened room. Making a small hole in his window shutter, he produced a thin beam of light which he then passed through a specially shaped piece of glass called a prism. The light emerged from the prism as a band of colours or **spectrum**.

Newton then carried out a further experiment placing a second identical but inverted prism behind the first. The coloured lights from the first prism were made to recombine by the second, producing white light. Newton concluded that white light is made up of the coloured lights of the spectrum.

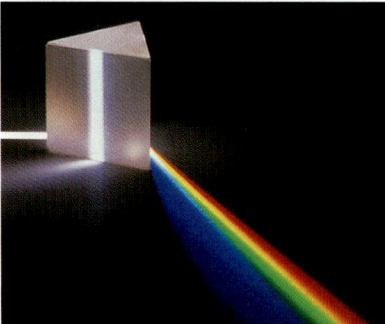

Figure 7 ▲ White light is made up of a spectrum of seven colours

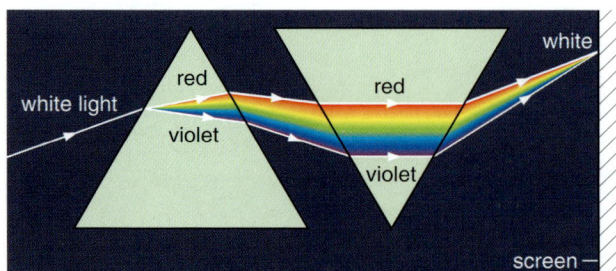

Figure 8 ▲ A second prism recombines the colours of the spectrum to produce white light again

A spectrum of white light is created because the different colours from which it is made are refracted by different amounts. This effect is called **dispersion**. Each of the different colours travels through the glass at a slightly different speed. The red light is travelling fastest through the glass, the violet is travelling most slowly.

Rainbows

Newton also realised that his theory could explain how rainbows are produced. He suggested that light is dispersed as it enters and leaves a water droplet, creating the spectrum of colour we call a rainbow.

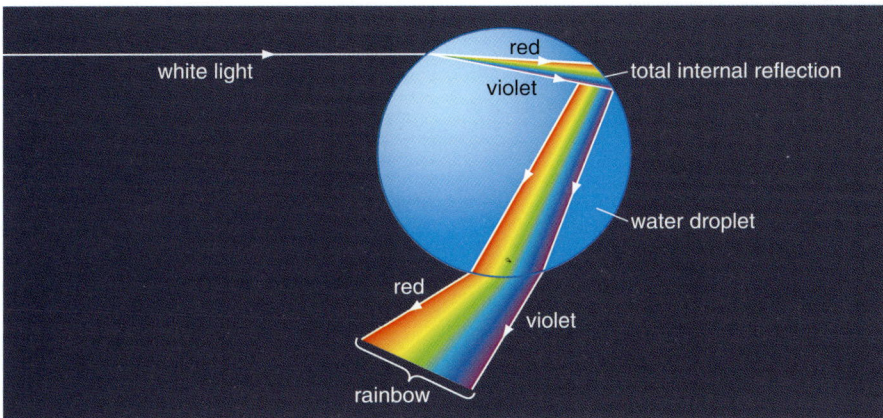

Figure 9 ◄ How a water droplet produces a spectrum of colour

The colours in the spectrum produced by a single prism and by a water droplet are the same. In order they are:

- red
- orange
- yellow
- green
- blue
- indigo
- violet

This sequence is easily memorised using the mnemonic **R**ichard **O**f **Y**ork **G**ave **B**attle **I**n **V**ain.

Test Yourself

4   Explain what happens to a ray of white light as it passes through a prism. Include a diagram in your answer.

5   What conclusion did Sir Isaac Newton come to when he investigated the coloured lights produced by a prism?

6   What causes the dispersion that produces a rainbow?

7   Sometimes a second rainbow can be seen in the sky. Find out how this secondary rainbow is produced.

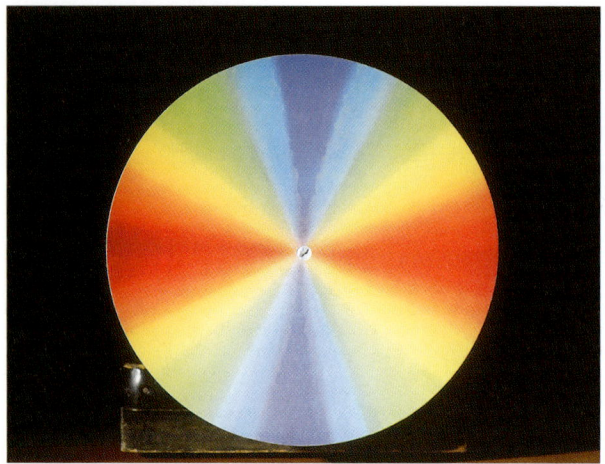

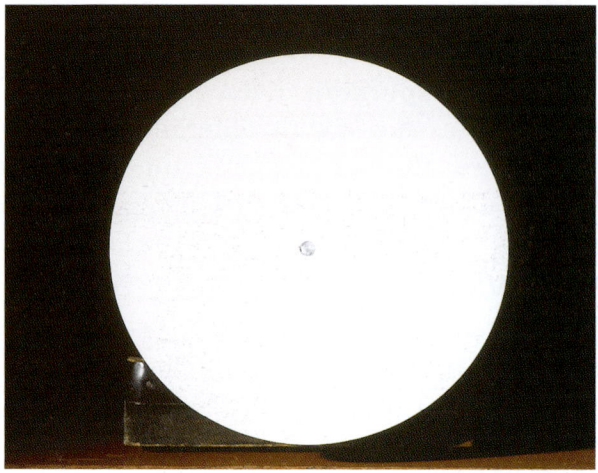

Figure 10 ▲ A Newton's disc a) stationary and b) spinning

Newton's disc

If this disc is spun quickly, the colours appear to mix and its surface appears to be white. This again confirms that white light is a mixture of coloured lights.

Coloured objects

Most of the objects you can see around you are coloured. They have a colour because they contain a chemical called a **dye** or pigment. When white light strikes a coloured object, all the colours in the white light are absorbed by the object except for those colours that are in the dye. These colours are reflected.

Figure 11 ▼ a) When white light strikes a blue object, all the colours are absorbed except for blue. This is reflected, so the object looks blue. b) When white light strikes a green object, all the colours are absorbed except for green. This is reflected, so the object looks green. c) When white light strikes a red object, all the colours are absorbed except for red. This is reflected, so the object looks red. d) When white light strikes a white object, none of the colours are absorbed. All the colours are reflected, so the object looks white. e) When white light strikes a black object, all the colours are absorbed. None are reflected, so the object looks black

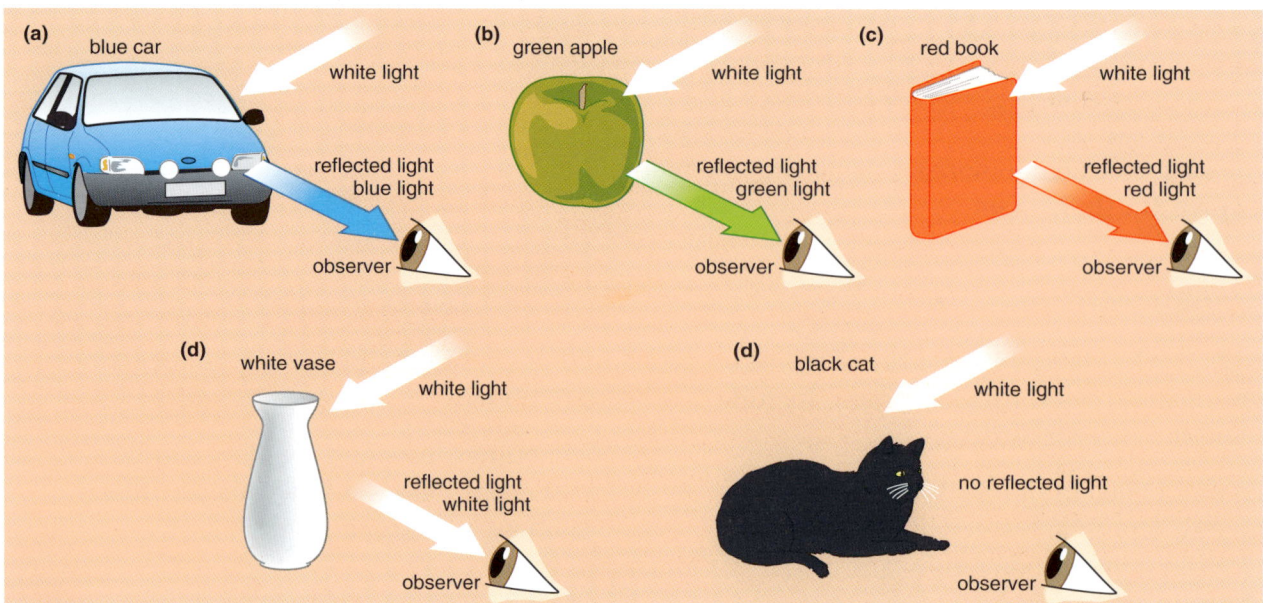

Adding coloured lights

If we shine coloured lights onto a screen where they overlap, we see new colours.

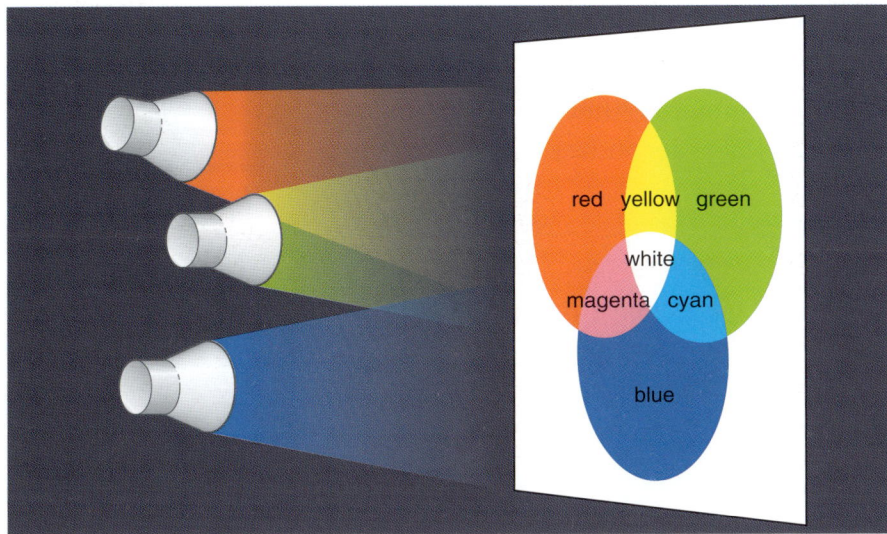

Figure 12 ◄ New colours are formed when coloured lights are mixed

The mixing of lights in this way is very useful in theatres where different coloured spotlights can be shone onto an object or person in order to give them a particular colour or create a particular atmosphere. There are, however, three colours that can not be made by mixing other coloured lights. These are red, green and blue. They are known as the **primary colours**.

Colour television

A colour television contains three electron guns. One gun produces a red spot on the screen, the second a blue spot and the third a green spot. The guns are controlled by the signals received through the television aerial. When a red colour signal is received, the red gun is aimed at a particular point on the screen. Electrons released by the gun cause this spot to glow red. The image we see on the screen is created by these glowing spots. Different combinations of the glowing coloured spots produce images of different colours. For example, equal numbers of red and green glowing dots will produce a region of yellow on the screen.

Extension box

Artists produce the colours they require by mixing different paints. There are three colours that an artist cannot make by mixing – these are red, blue and yellow. In art, therefore, red, blue and yellow are known as the primary colours.

Figure 12 shows the effect of mixing together the lights that are primary colours. The results of this simple experiment can be summarised more usefully in a **colour triangle**.

The triangle tells us that:

- If the three primary colours are mixed in equal amounts, white light is produced.

- If two primary colours are mixed in equal amounts, they produce a **secondary colour**. There are three secondary colours – yellow, cyan and magenta.

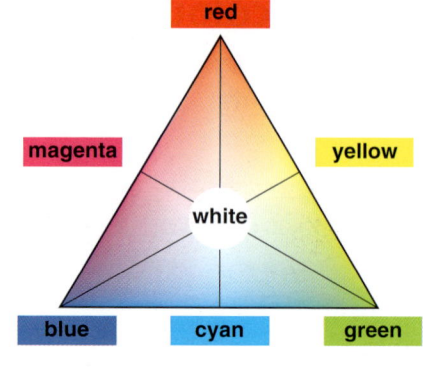

Figure 13 ▲ The colour triangle

Test Yourself

8   Using the colour triangle, copy and complete Table 1 below:

Colour 1	Colour 2	Colour 1+ Colour 2
red	green	
blue	red	
yellow	blue	
blue		cyan
magenta		white

Table 1 ▲

Extension box

Looking at coloured objects under coloured lights

Identifying the colour of a car in daylight is no problem at all to most of us but identifying the colour of a car at night under street lights is not so easy.

We see non-luminous objects because of the light they reflect. We see coloured objects in daylight because they absorb some colours of light and reflect others. But what happens if the light shining on an object is not white? Will this make the object look a different colour? To solve this problem you need to ask yourself the question does the light shining on the object contain light which is of the same colour as the object? If the answer is yes, the object will look its normal colour.

Example

A green apple will appear green in a) white light, green light, yellow light and cyan light because all these lights contain green.

Extension box

If, however, red light is shone onto the green apple, it is absorbed. There is no green light to be reflected so the apple looks black. Similarly if blue light is shone onto the apple, it will look black.

Test Yourself

9 Using the colour triangle, work out the apparent colour of each object to copy and complete Table 2 below.

Colour of light shining on object	Real colour of object	Apparent colour of object
red	red	
blue	white	
green	red	
yellow	green	
magenta	red	
red	magenta	
yellow	blue	
cyan	green	
red	yellow	
red	cyan	

Table 2 ▲

10 Most street lights are yellow, so what colours of cars will still be seen in their correct colour and what colours of cars will appear black?

Filters

A **filter** is a coloured piece of glass or plastic. When light is shone through a filter some colours of light are absorbed.

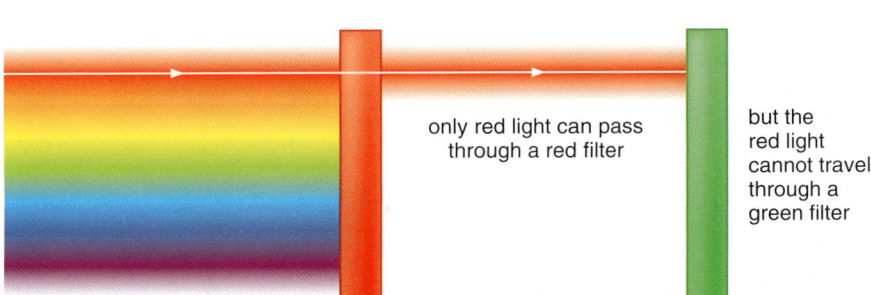

only red light can pass through a red filter

but the red light cannot travel through a green filter

Figure 14 ◄ The red filter will only allow red light to pass through it. The blue light is absorbed. The green filter will only allow green light to pass through it. The red light is, therefore, absorbed

95

Test Yourself

11 Using the colour triangle, work out the colour of the light allowed through each of the filters in Table 3 below:

Colour of light	Colour of filter	Colour of light allowed through filter
white	red	
blue	green	
yellow	red	
yellow	cyan	
magenta	green	
cyan	magenta	

Table 3 ▲

Summary

When you have finished studying this chapter, you should understand that:

✔ Rays of light travel at different speeds in different media.

✔ As a ray crosses from one medium to another, the change in speed may cause it to change direction. This effect is called refraction.

✔ As a ray of light enters a more dense medium, it slows down and bends towards the normal.

✔ As a ray of light enters a less dense medium, it increases its speed and bends away from the normal.

✔ White light is a mixture of coloured lights.

✔ Different coloured lights are refracted by different amounts. This effect is called dispersion.

✔ Coloured objects reflect light of their own colour.

✔ There are three primary colours – red, green and blue. All other colours can be made by mixing primary colours.

✔ Filters are transparent to some colours but absorb others.

End-of-Chapter Questions

1 Explain in your own words the following key terms you have met in this chapter:

medium	dye
refraction	primary colour
normal	colour triangle
spectrum	secondary colour
dispersion	colour filter

2 Read the question and answer below.

Question What colour will a green object look under yellow light?

Answer Yellow light is a mixture of red and green light. When the yellow light strikes the green object, the red light is absorbed and the green light is reflected so the object looks green.

Now write a similar explanation for the first five objects in Table 2 on page 95.

3 Read the question and answer below:

Question When yellow light is shone onto a magenta filter, what colour(s) of light pass through the filter?

Answer Yellow light is a mixture of red and green light. Magenta is a mixture of blue and red light. When the yellow light strikes the magenta filter, the green light is absorbed and the red light passes through.

Now write a similar explanation for the filters in Table 3 on page 96.

4 Take a coloured picture from a magazine and look at it closely using a magnifying glass or a low-powered microscope. Explain in your own words how the colours in the picture are produced.

5 Find out what is meant by being 'colour blind' and how it is tested.

10

Sound and hearing

Producing sounds

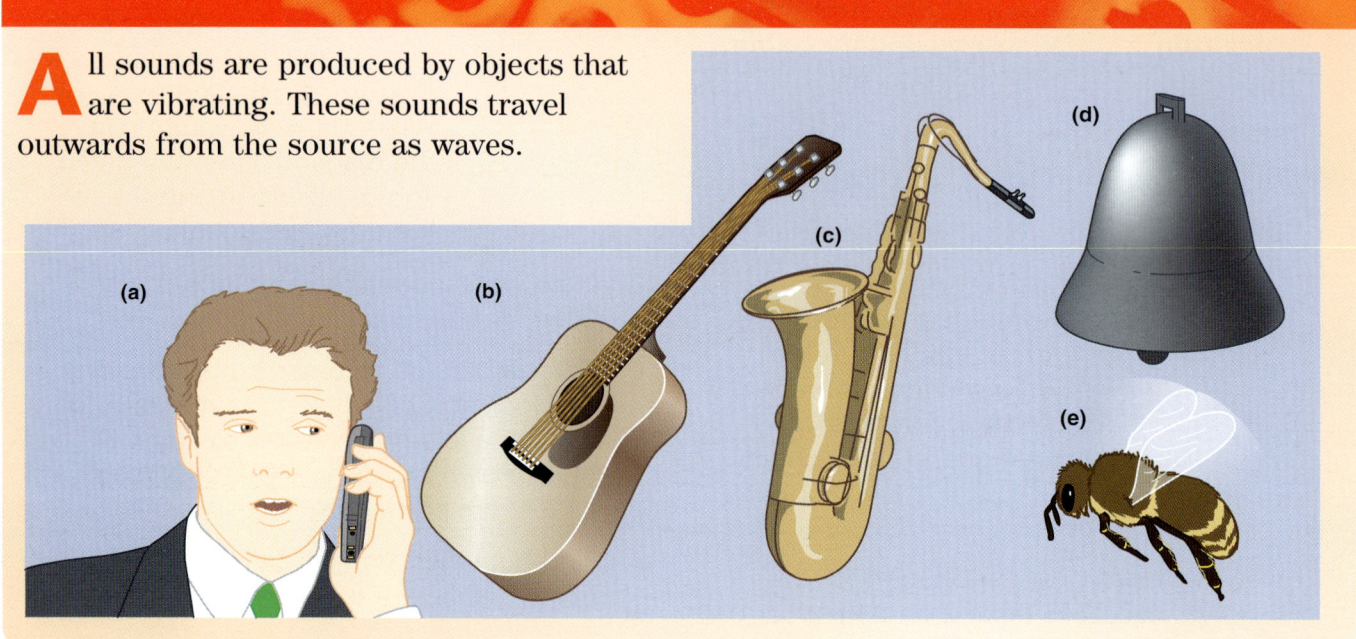

All sounds are produced by objects that are vibrating. These sounds travel outwards from the source as waves.

Figure 1 ▲ a) When we speak, our vocal cords vibrate, producing sounds

b) When the strings of this guitar are plucked, they vibrate, producing musical sounds

c) Musical sounds are also produced by instruments like this saxophone. The air inside the saxophone is made to vibrate by blowing

d) When the clanger strikes the bell, the vibrations produced are heard as a ringing sound

e) The buzzing sound we hear from a bee is produced by its vibrating wings

Test Yourself

1 Name three different musical instruments and explain how each of them produces sounds.

Creating and hearing sound waves

Sound waves travel from vibrating objects to our ears by means of sound waves. Figure 2 shows how a sound wave is created.

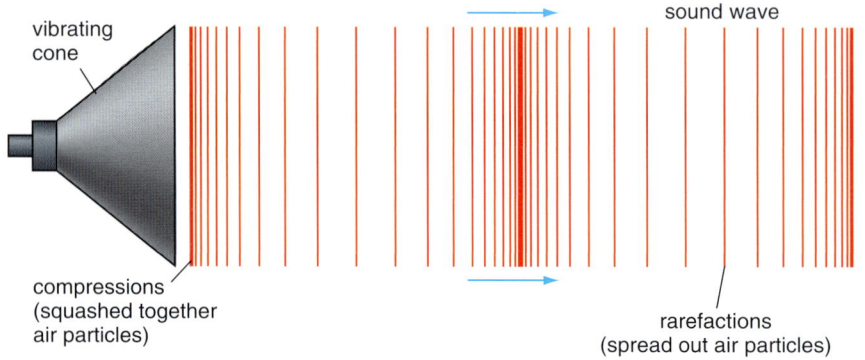

vibrating cone

sound wave

compressions (squashed together air particles)

rarefactions (spread out air particles)

Figure 2 ◄ A sound wave

As the object vibrates to the right, it pushes the air particles close together, creating a **compression**.

As the object vibrates in the opposite direction, a region of *spread-out* air particles is created behind it. This is called a **rarefaction**.

When the object vibrates to the right again, another compression is created. After several vibrations a series of compressions and rarefactions have been created that are moving away from the object. This is a sound wave.

Test Yourself

2 What is a compression?

3 What is a rarefaction?

4 Draw a diagram to explain how a vibrating object creates a sound wave in air.

The ear

Figure 3 ▼ The structure of the ear

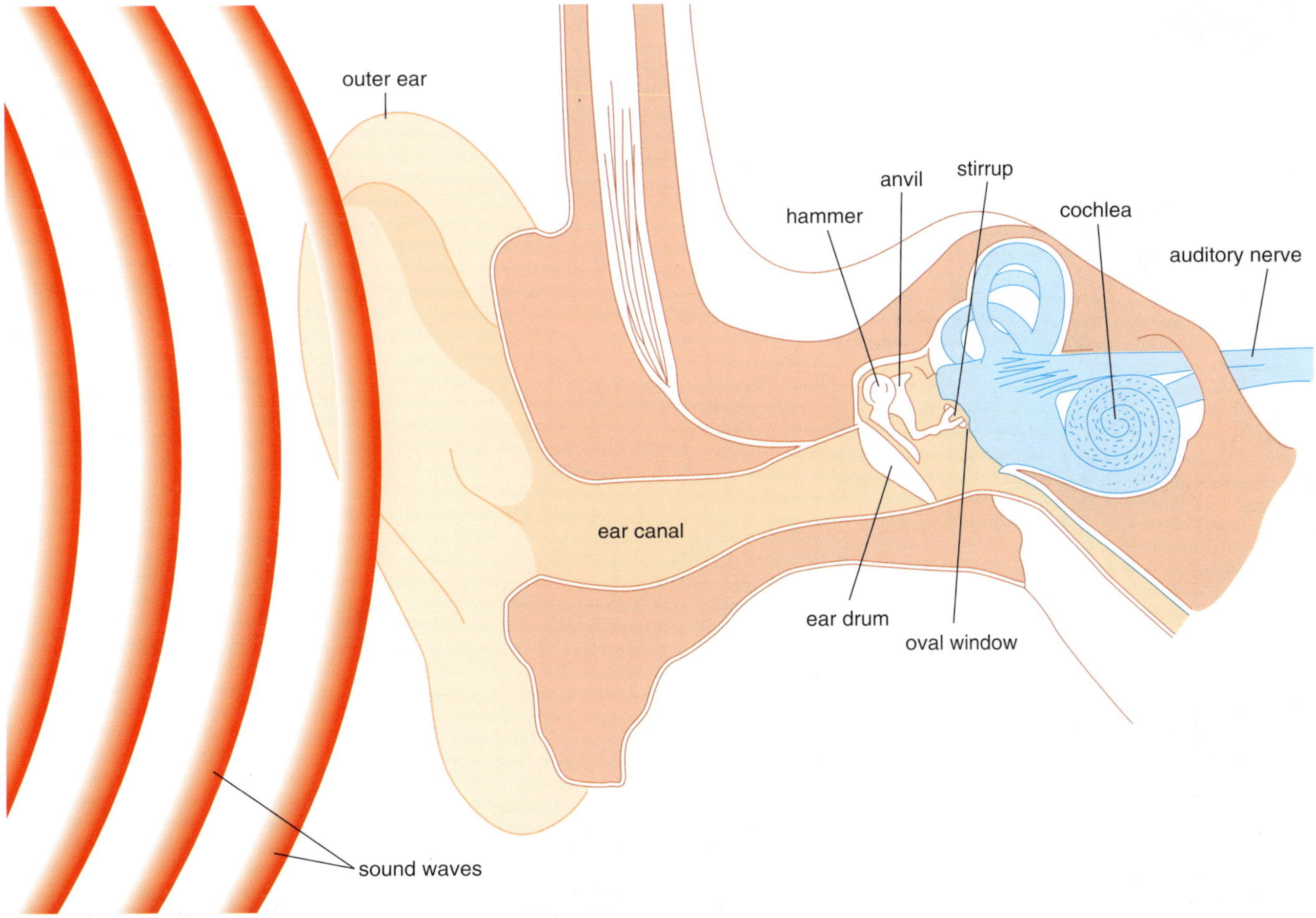

outer ear

hammer

anvil

stirrup

cochlea

auditory nerve

ear canal

ear drum

oval window

sound waves

Sound waves are 'collected' and directed into the ear canal by the outer ear. When the sound waves strike the **eardrum**, it is made to vibrate. The eardrum is connected to three tiny bones called the **hammer**, the **anvil** and the **stirrup**. As the eardrum vibrates, these three bones behave like levers and amplify the vibrations so that the ear can work better. These vibrations then pass through a liquid inside a long coiled structure called the **cochlea**. **Hair cells** along the inside of the cochlea are stimulated by the vibrations travelling through the liquid. Impulses are sent from the hairs to the brain through the auditory nerve. The brain then 'hears' the sound.

Sound waves must have something to travel through

Figure 4 ◄

Sound waves can travel through solids. That's why people can hear your music through the walls and floor of your room if you have it turned up too loud. Sound waves can also travel through liquids. That is why whales, dolphins and some fish are able to communicate with each other whilst swimming deep under the ocean surface. Sound waves can also travel through gases such as air. That is how we are able to hear all the sounds around us.

Test Yourself

5　What is the purpose of the outer ear?

6　What do the three bones in the centre of the ear do?

7　What do the hair cells inside the cochlea do?

For sounds to be able to travel, there must be particles. The bell jar experiment shown in Figure 5 demonstrates that sounds cannot travel through a space that has no particles i.e. a **vacuum**.

When there is air in this bell jar, we are able to see and hear the bell ringing. If the air is pumped out of the jar and the bell turned on, we can see it ringing but we cannot hear it. There are no particles in the jar to carry the sound energy.

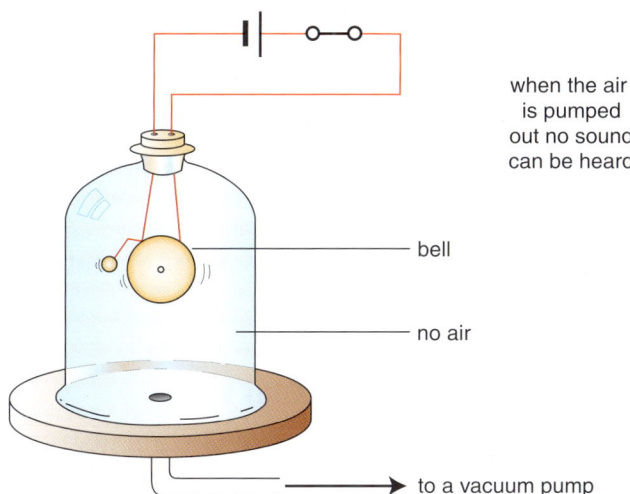

when the air is pumped out no sound can be heard

bell

no air

to a vacuum pump

Figure 5 ▲ The bell jar experiment shows us that sound cannot travel through a vacuum

Test Yourself

8   Explain in your own words why the bell jar experiment proves that light waves can travel through a vacuum and sound waves cannot.

Frequency of vibration and pitch

If we place a metre ruler over the edge of a desk and twang it to make it vibrate, we can watch the vibrations and hear the sounds the ruler is producing. If the length of the ruler which is over the edge of the desk is decreased and then again twanged, we see a difference in the vibrations of the ruler and hear a difference in the sounds it is producing. The vibrations are quicker and the sounds produced have a higher **pitch**.

A similar experiment with tuning forks of different sizes produces the same result. The longer tuning fork vibrates more slowly. It has a low frequency and produces a low pitched sound. The shorter tuning fork vibrates more rapidly. It has a higher frequency and produces a higher pitched sound.

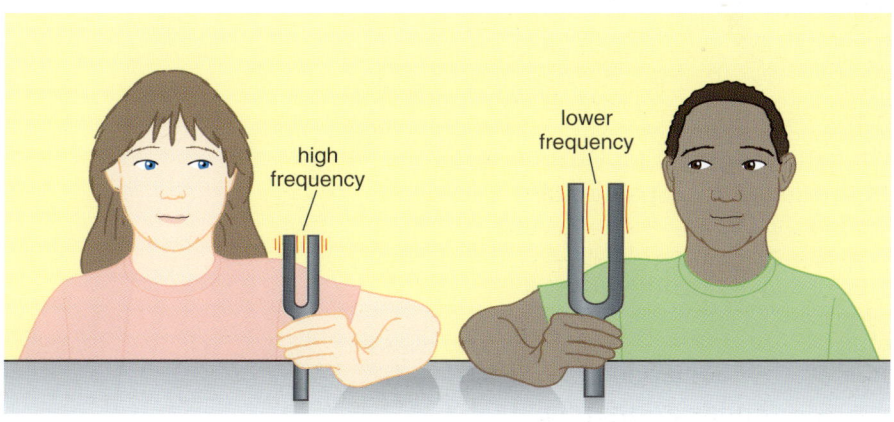

high frequency

lower frequency

Figure 6 ◄ Tuning forks of different sizes produce notes of different frequencies when they vibrate

The **frequency of an object/wave** is the number of complete vibrations it performs each second. It is measured in **hertz** (Hz), where 1 Hz is one vibration per second.

> **Small objects vibrate quickly and produce high pitched sounds.**
>
> **Large objects vibrate slowly and produce low pitched sounds.**

Stringed instruments

A double bass is a much larger musical instrument than a violin and can therefore produce much lower notes. Both instruments can produce notes with a wide range of frequencies and pitch. The changes in the frequencies produced by vibrating strings are achieved by altering:

- The *length* of the string. The longer the string, the slower the vibration and the lower the note produced.

- The *thickness* of the string. The thicker the string, the lower the pitch of the note produced.

- The *tension* of the string. The greater the tension in the string, the higher the pitch of the note produced.

Wind and brass instruments

Changes in the frequencies of the sounds produced by wind and brass instruments are achieved by altering the lengths of the columns of air which vibrate when they are played.

The more recorder holes that are covered with the fingers, the longer the vibrating air column and the lower the note produced by the recorder.

As a trombone slide is pushed out, the air column inside becomes longer and the pitch of the notes heard becomes lower.

Seeing sound waves

It is not possible for us to see sound waves, but using a piece of apparatus called an **oscilloscope**, we can produce a picture on a screen to represent the sound waves produced by a particular source.

The oscilloscope shows us the number of waves produced in a certain time.

Test Yourself

9   Explain why the short strings of a harp produce high pitched notes and the longer strings produce lower pitched notes.

10   Explain how a trombone player changes the notes produced by his trombone.

11   Explain how a flying mosquito produces a sound. Why is this sound high pitched?

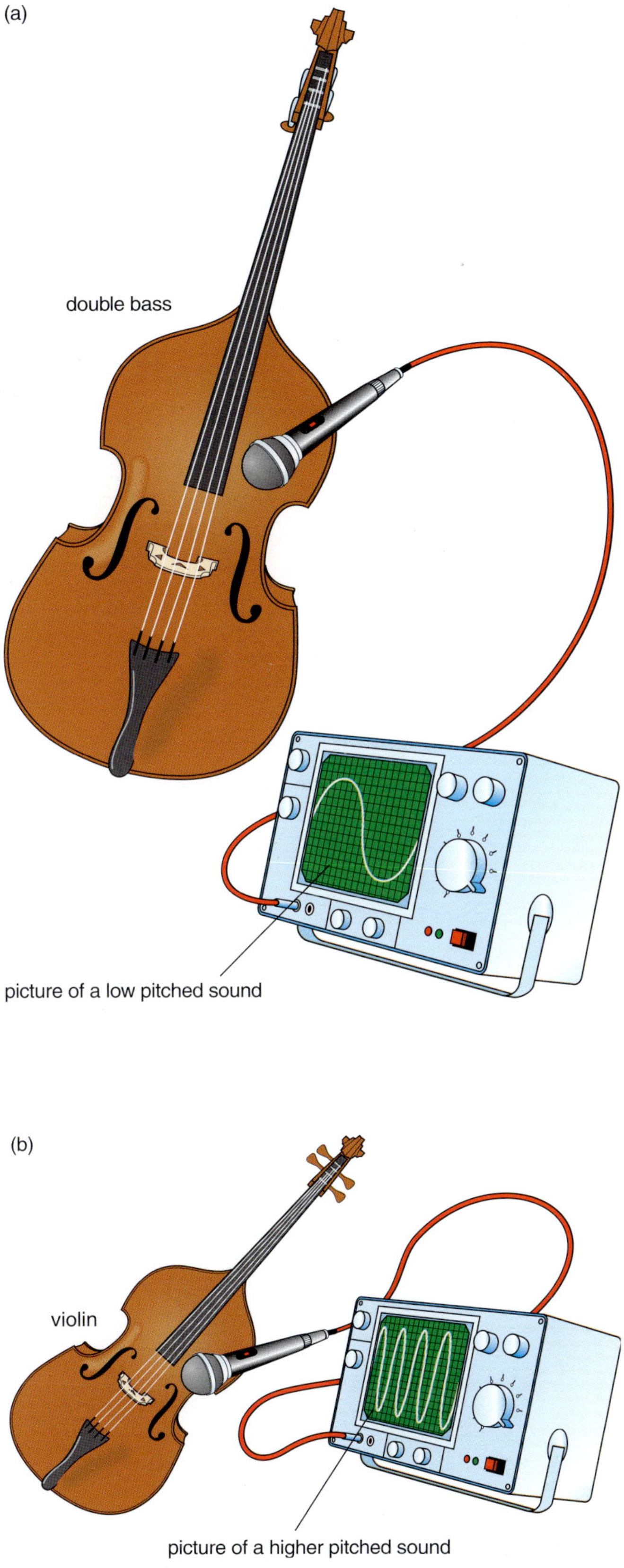

(a)

double bass

picture of a low pitched sound

Loudness and amplitude of vibration

Loud sounds are seen on the oscilloscope as tall waves. Tall waves are said to have a large **amplitude**. Quiet sounds are seen as waves with small amplitudes.

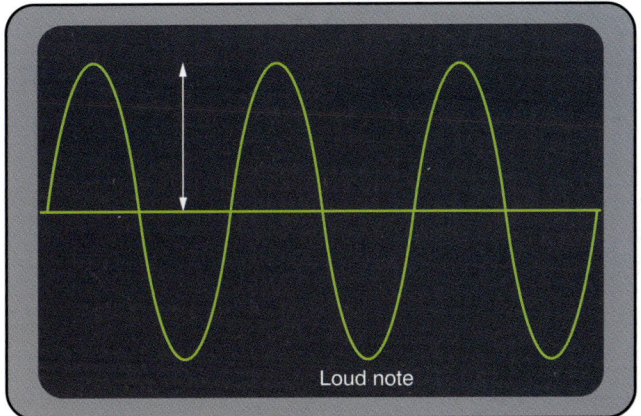

Loud note

Quiet note

Figure 8 ▲   Quiet sounds have a small amplitude, while loud sounds have a large amplitude

(b)

violin

picture of a higher pitched sound

Figure 7 ▲   Oscilloscope displaying a) low pitched sound and b) higher pitched sound.

Tone

If we closed our eyes and listened to someone playing a violin, trombone and guitar, we would have no difficulty in deciding which instrument made which sound. This is because the sounds have different **tones or quality**. Notes of different tones are shown on the oscilloscope as waves with different shapes.

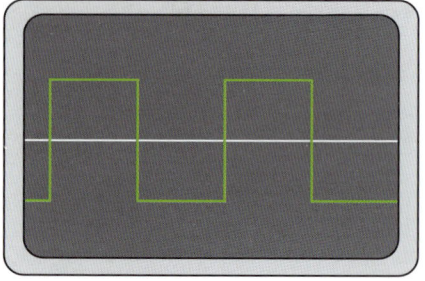 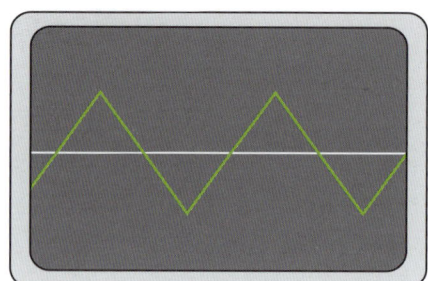

Figure 9 ◄ Different instruments produced different shaped waves on the screen

Loudness

We hear sounds when air particles strike our eardrums making them vibrate. If lots of air molecules strike our eardrums and they are carrying a lot of sound energy, the sound we hear is loud. If only a few particles strike our eardrums, there is little sound energy and the sound we hear is soft or quiet.

If an object vibrates with a large amplitude, it will produce loud sounds. If it vibrates with a smaller amplitude, it will produce quieter sounds. For example, if a drum is struck very hard, the skin vibrates with a large amplitude and creates a loud sound. Conversely, if the drum is struck softly, the skin vibrates with a small amplitude, creating a quiet sound.

Very loud sounds can cause damage to your hearing. Even listening to your personal stereo on a loud setting over a long period of time can permanently damage your hearing. People who have to work in very loud environments, such as airports or building sites, often wear ear defenders to protect their ears. It is, therefore, important that we monitor sound levels in order to avoid these problems. We measure the loudness of a sound on the **decibel scale**. Table 1 will give you some ideas of the values of some sounds on the decibel scale.

Type of sound	Average loudness (dB)
silence	0
a whisper	25
normal conversation	60
noisy traffic	80
noisy factory	90
noisy disco	110
jet taking off	125
sounds that cause physical pain	above 130

Table 1 ▲

Test Yourself

12 Draw an oscilloscope diagram of a wave for a sound which is:
   a) high pitched and loud
   b) low pitched and quiet.

13 Draw oscilloscope diagrams of waves for two sounds of the same amplitude and pitch but with different tones.

Test Yourself

14 Explain the difference in the vibrations of a drum skin when it is producing a) a loud sound and b) a quiet sound.

15 What is the decibel scale?

16 What are ear defenders and where are they used?

Noise and pollution

Unwanted sound is called **noise**. Noise pollution can be a real problem. It can cause stress and make it more difficult to concentrate. There are several ways in which planners, builders, architects and engineers try to reduce this problem. They may:

- put barriers such as walls or rows of trees between the source of sound and nearby buildings;

- double glaze buildings close to the source of the sound e.g. double glaze the windows of homes near an airport;

- develop machinery that is quieter, such as better insulated car engines.

Hearing range

Some objects vibrate so quickly that we are unable to hear the sounds they produce. These sounds are called **ultrasounds**. Although humans are unable to hear these sounds, some animals can. Bats hunt and communicate with each other using ultrasounds. Dog whistles produce ultrasounds that can be heard by dogs but not humans.

Figure 10 ◄ A bat uses ultra sounds to help it find and catch insects

Some objects vibrate so slowly they also produce sounds we cannot hear. For example, many of the sounds produced by elephants cannot be heard by humans.

Science Scope: PHYSICS

The range of frequencies that can be heard by humans varies a little from person to person. In general the **hearing range** or audible range of a human is from 20 Hz to 20 000 Hz. Older people usually have a narrower hearing range than the young.

Ultrasounds have many uses. They are used in hospitals to scan a foetus in the womb. Reflected ultrasonic waves are detected by a probe and these signals are processed by a computer which produces an image of the foetus on a screen.

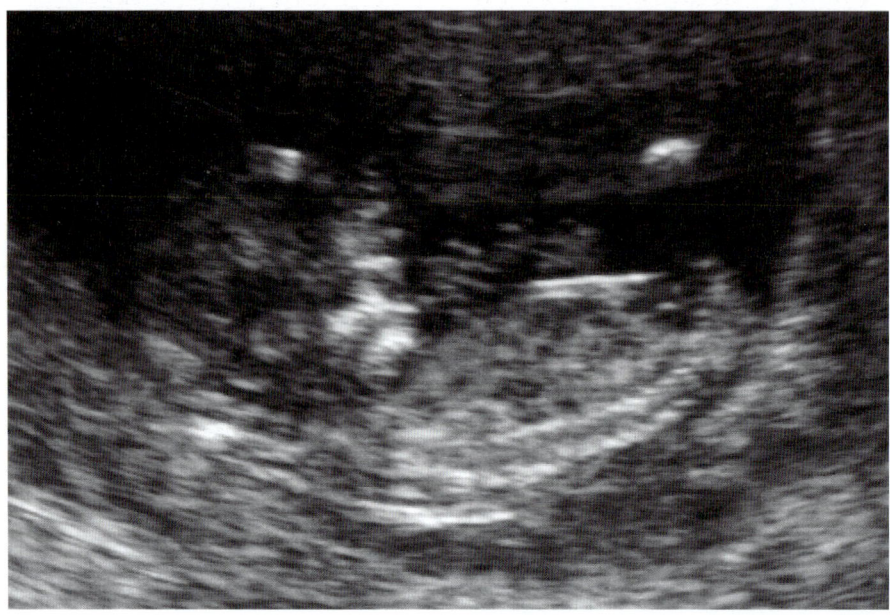

Figure 11 ◄ Doctors use ultrasound scans to make sure unborn babies are healthy

Test Yourself

17 What is noise?

18 State three ways in which the effects of noise could be reduced?

19 What are ultrasounds? Give one use of ultrasounds.

Speed of sound

The speed of sound in air is approximately 340 m/s or 760 mph. (This value changes if the temperature changes. Sound waves travel faster in warm air than in cold air.) The speed of sound is very much slower than the speed at which light travels. Light travels through air at 300 000 000 m/s. It is this large difference in their speeds that sometimes results in an action being seen before it is heard e.g. thunder and lightning, or seeing a cricketer play a shot before hearing it etc. (see page 155).

Sounds travel at different speeds in different media. As a general rule they travel fastest in solids and slowest in gases.

Medium	Speed of sound
air	340 m/s
water	1500 m/s
concrete	5000 m/s
steel	6000 m/s

Table 2 ▲ The speed of sound in various different media

Extension box

It was thought for a long time that it was impossible for humans to travel faster than the speed of sound. But on 14th October 1947, an American airforce officer named Chuck Yeager flew his Bell X-1 rocket powered plane through the sound barrier at a speed of 1126 kph. Aircraft that travel faster than the speed of sound are called **supersonic**. As an aircraft exceeds the speed of sound it passes through its own sound, creating a shock wave which spreads out behind it. We hear this shock wave as a **sonic boom**.

Figure 12 ▲   The Bell X-1

Summary

When you have finished studying this chapter, you should understand that:

✔ Sounds are produced by objects vibrating.

✔ Sound energy travels in the form of a wave.

✔ Sound waves can travel through solids, liquids and gases but not through a vacuum.

✔ Large objects vibrate slowly, producing low pitched notes. Small objects vibrate quickly and produce high pitched notes.

✔ We can 'see' sound waves by using an oscilloscope.

✔ The larger the amplitude of vibration, the louder the sound the object produces.

✔ Loudness of sounds is measured on the decibel scale. Constant exposure to loud sounds can permanently damage our hearing.

✔ Unwanted sounds are called noise. Controlling noise pollution is important for our well being.

✔ Human beings have a hearing range of 20 Hz to 20 000 Hz. Sounds with frequencies higher than this are called ultrasounds.

✔ Sound waves travel at different speeds in different media.

End-of-Chapter Questions

1 Explain in your own words the following key terms you have met in this chapter:

compression   peak of a wave

rarefaction   wavelength

eardrum   amplitude of
vibration

hammer, anvil and
stirrup   tone or quality

cochlea   decibel scale

hair cells   noise

vacuum   ultrasound

pitch   hearing range

oscilloscope   **supersonic**

wave frequency **sonic boom**

hertz

2 The diagram below shows the picture of a sound wave produced by an oscilloscope.

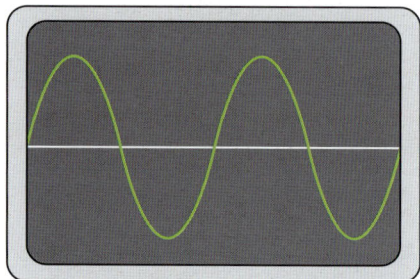

a) Copy this diagram into your books and label the amplitude of the wave.

b) Draw a second diagram to show what would happen if the frequency of the source of this wave increased.

c) Draw a third diagram to show what would happen to the wave in part a) if the source produces a louder sound.

d) How would the wave change if the tone of the sounds from the source is altered?

3 What is the decibel scale? Suggest values on the decibel scale for the following:

a) someone shouting

b) someone whispering

c) popping a balloon

d) a racing car going past

e) a noisy engineering factory.

4 What is an echo? Find out how you could use echoes to determine the speed of sound in air.

5 The diagram below shows two pupils using a string telephone.

a) What is the source of the sound?

b) Explain in detail how the sounds travel to the listener's ear.

c) Explain what happens to the sound energy when it reaches the listener's ear.

11 The Earth in space

The Earth

We live on a **planet** called the Earth. If we could see our planet from afar it would look like Figure 1.

Figure 1 ▼ The Earth from space

The Earth is a rocky planet but more than two-thirds of its surface is covered with water.

Although as we go about our everyday lives we cannot feel it, the Earth is spinning around like a top. It completes one full turn or rotation every 24 hours. This is one Earth **day**.

It is this turning motion which makes the Sun appear to rise in the East, travel high in the sky and then set in the West.

Although the path followed by the Sun is always from East to West, it does change with the seasons. In the summer months the Sun follows a much higher path than it does during the winter.

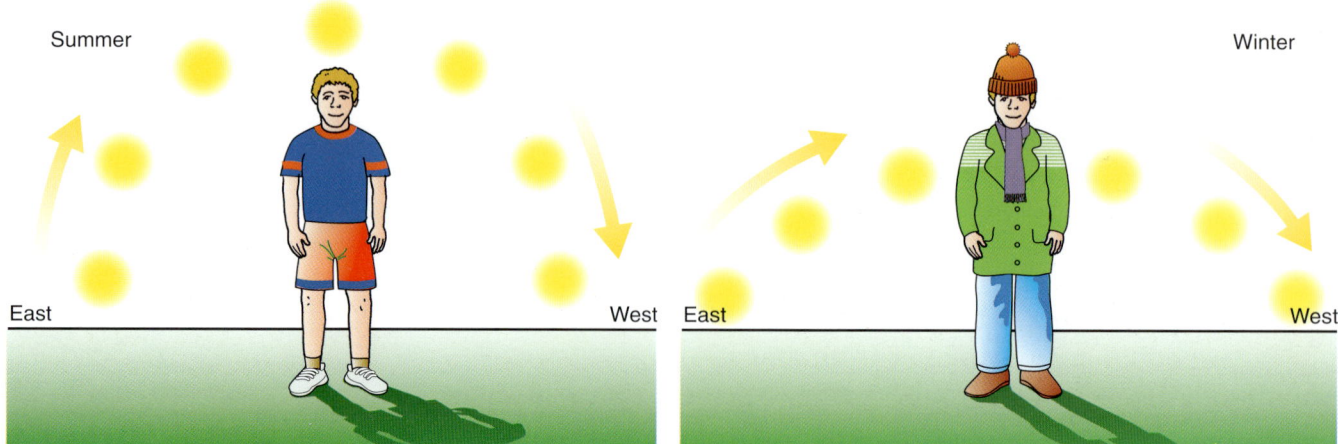

Figure 2 ▲   The path of the Sun across the sky is lower in the winter and higher in summer

Day and night

The Earth's turning motion also gives rise to day and night. As it turns, part of the Earth's surface faces the Sun whilst other parts are in darkness. It is **daytime** on those parts of the Earth that are receiving light from the Sun and **night-time** on those parts that are not receiving light from the Sun.

The Earth spins from West to East. This means that places that are to the East of you move into the light or the shadow before you do. Places that are to the West of you move into light and shadow after you.

People living at A in Figure 4 are in the middle of the shadow region i.e. it is in the middle of their night. People living at B are just entering the shadow region i.e. it is sunset here. People living at C are in the middle of the sunlit region i.e. it is midday here.

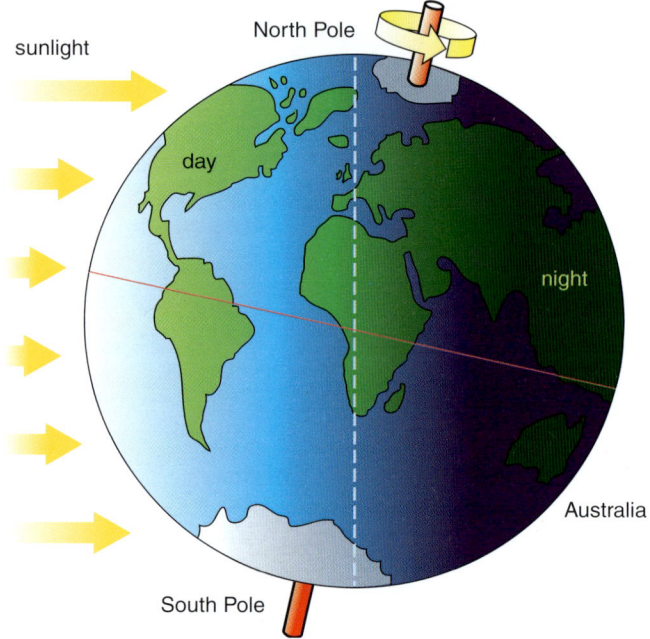

Figure 3 ▲   How the spinning of the Earth causes day and night

Test Yourself

1   How long will it take for the Earth to make seven full turns about its axis?

2   Draw a diagram to show the paths taken by the Sun on a) a summer day and b) a winter day.

3   Explain when it is
    a) night-time and
    b) daytime in Britain.

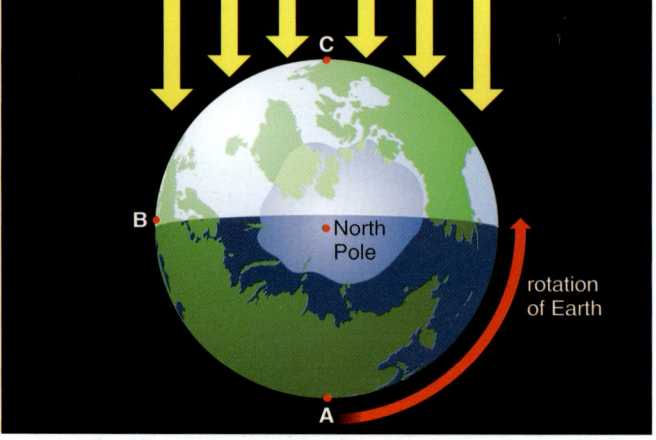

Figure 4 ▲

Extension box

Time zones

Sunlight arriving at different times in different parts of the world used to create real problems for governments all over the world. For example, sunlight arrives on the east coast of England several minutes before it reaches parts of Cornwall. Should the clocks and watches in these places be showing different times? In 1884, the governments agreed to divide the world into 24 **time zones**. Within any one time zone all the clocks would show the same time.

In the time zone to the east, the clocks would be set one hour earlier and in the time zone to the west, they would be set one hour later. The diagram below shows the number of hours a time zone is in front or behind **Greenwich Mean Time** (the time in London). On the far side of the Earth from London a line is drawn from the North pole to the South pole. It is called the **international date line** and indicates where one day ends and another begins.

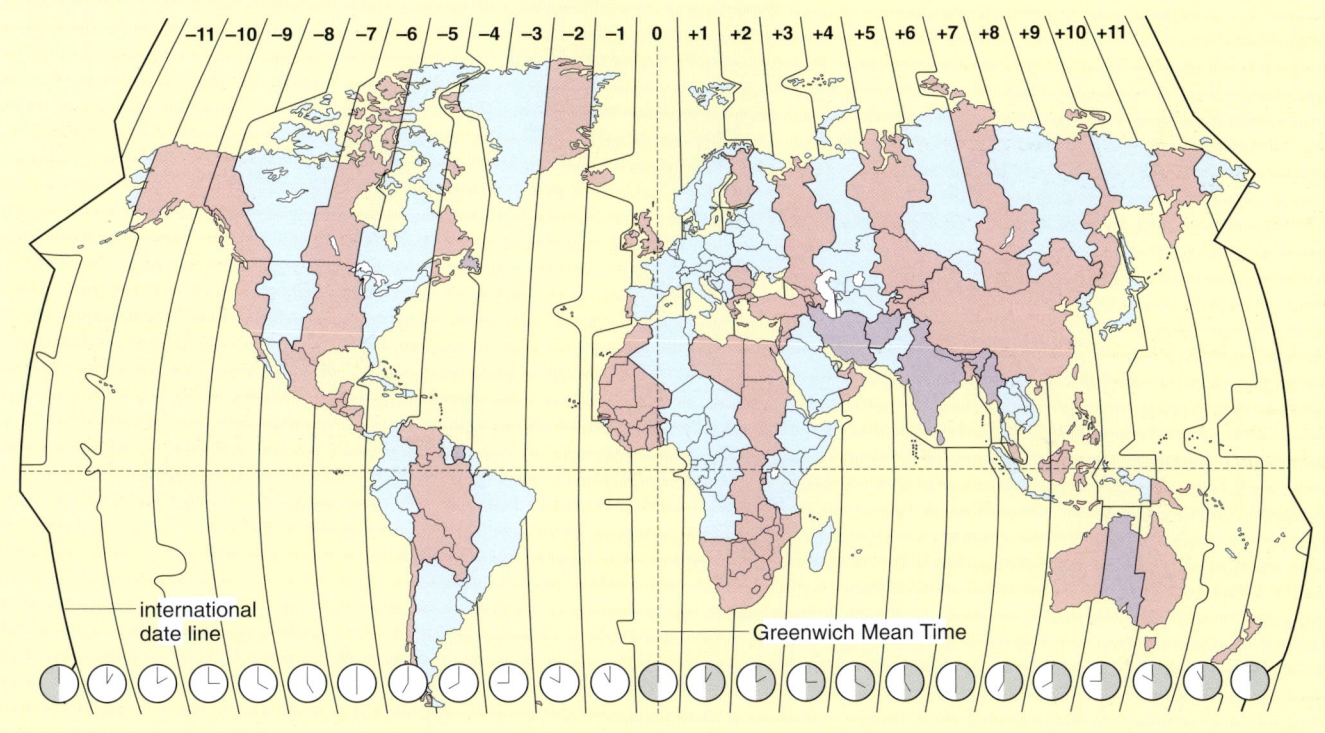

Figure 5 ▲ How the world is divided into time zones

A year and the seasons

The nearest star to the Earth is the Sun. It is 150 million kilometres away. The Earth orbits the Sun once every year, following a path called an **ellipse**. An ellipse is a slightly squashed circle. As the Earth travels around the Sun we experience the different **seasons**. We have seasons because the Earth's axis is tilted and this affects the amount of the Sun's energy reaching the different parts of the Earth's surface.

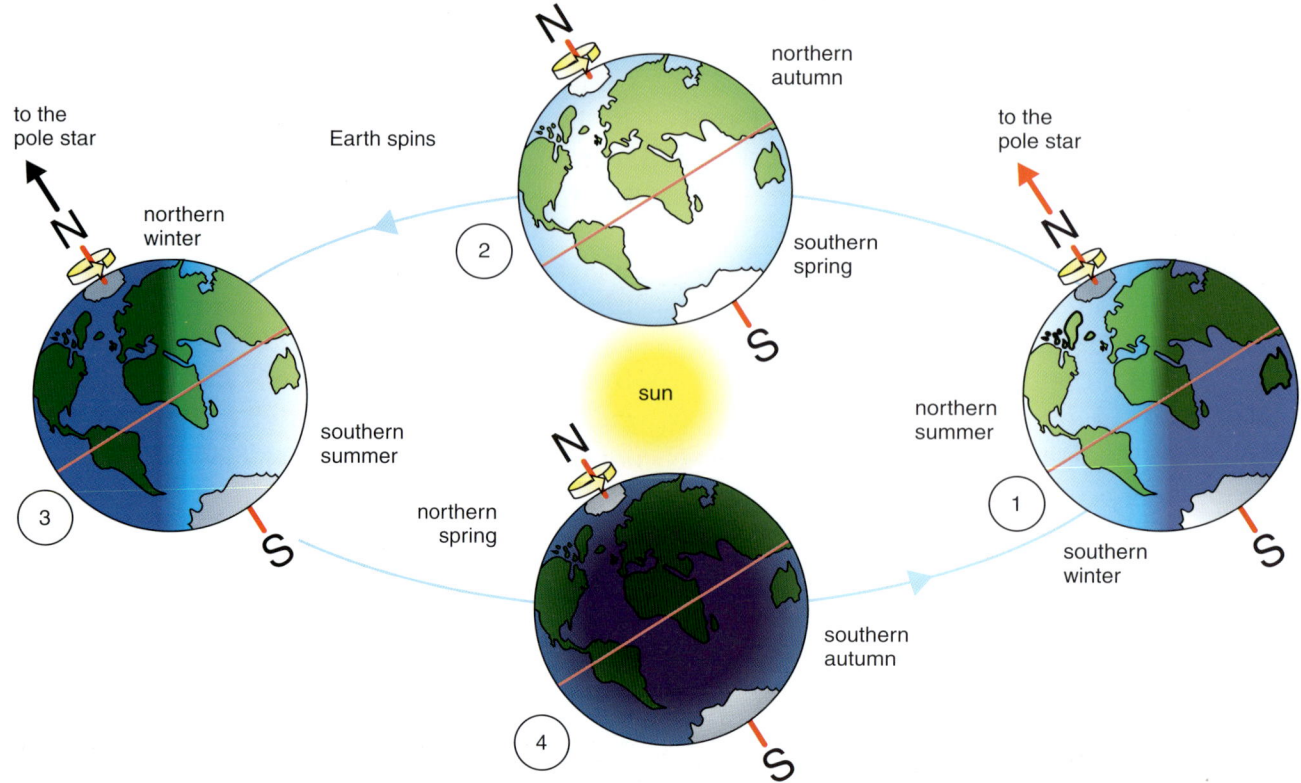

Figure 6 ▲   The tilting of the Earth causes us to have seasons

When we in Britain are in summer, our part of the Earth – the northern hemisphere – is tilted *towards* the Sun. This makes our weather warmer and our days longer. During this time we might receive as much as 16 hours of sunlight a day. When we are in winter, our part of the Earth is tilted *away from* the Sun. This makes our weather colder and our days shorter. During this time we might receive as little as 8 hours of sunlight a day.

Extension box

We usually say that the Earth orbits the Sun once every 365 days but this is not completely accurate. It actually takes the Earth 365¼ days to orbit the Sun. If no account was taken of this extra ¼ day, the seasons would gradually go out of step with the months and we could find ourselves celebrating Christmas in the summer! To avoid this problem we add an extra day, 29th February, to the year every 4 years. These years are called **leap years**.

The Solar System

The Earth is one of several planets that orbit the Sun. Altogether there are nine planets. Starting with the planet nearest the Sun, they are **M**ercury, **V**enus, **E**arth, **M**ars, **J**upiter, **S**aturn, **U**ranus, **N**eptune, **P**luto. An easy way to remember the names and the order is to use the sentence **M**any **V**ery **E**nergetic **M**en **J**og **S**lowly **U**pto **N**ewport **P**agnell.

The Sun, the planets and their moons make up the **Solar System**. The planets move in elliptical orbits with the Sun near the centre. The orbits of all the planets are in the same plane, except for that of Pluto. Pluto's orbit is at an angle to this plane.

The vast majority of the Solar System is empty space. It is impossible to draw a scale picture to show this. To get some idea of the scale, imagine you have a football and a pea. Put the football on the ground and then walk away from it. After 100 paces put the pea on the ground. You now have an approximate representation of the positions and sizes of the Sun and the Earth in the Solar System.

Test Yourself

4 How long does it take the Earth to orbit the Sun once?

5 Why do we have seasons on Earth?

6 What is a leap year?

Figure 7 ▼ The Solar System

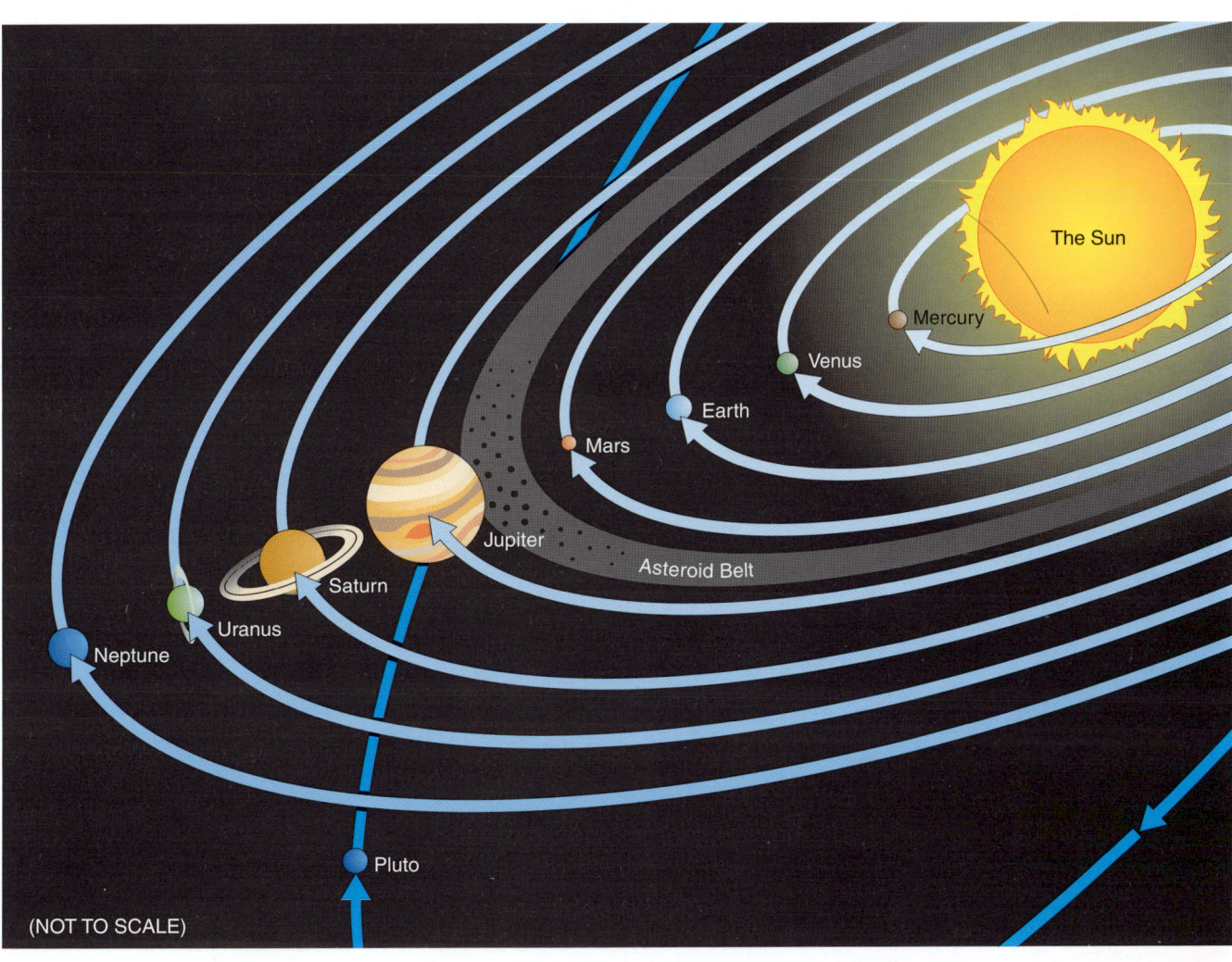

(NOT TO SCALE)

Information about the planets in our Solar System

Planet	Approximate distance from Sun compared with the Earth	Approximate diameter compared with the Earth	Approximate mass compared with the Earth	Approximate time to orbit the Sun in years	Approximate surface temperature	Nature of surface of planet
Mercury	0.5	0.5	0.05	0.5	430–180 °C	Rocky
Venus	0.75	1	0.8	0.5	480 °C max	Rocky
Earth	1	1	1	1	−55 – 35 °C	Rocky with large amount of water cover
Mars	1.5	0.5	0.1	2	Average 23 °C	Rocky
Jupiter	5	11	320	12	−160 °C	Swirling gases
Saturn	10	10	95	30	−150 °C	Swirling gases
Uranus	20	4	15	85	−220 °C	Gas
Neptune	30	3.5	318	165	−213 °C	Gas
Pluto	40	0.75	0.002	250	−230 °C	Rock and ice

Table 1 ▲

Our Solar System contains other bodies besides the Sun and the planets, such as moons, asteroids and comets.

Moons

Moons are natural objects which orbit a planet. The Earth has just one moon but Jupiter has 16 moons, Saturn has 23 moons and Mercury and Venus have no moons.

Figure 8 ▲ Our Moon

Phases of the Moon

The Moon, as we have already seen, is a **non-luminous** object. We see it because of the light it reflects. The amount of sunlit Moon surface that we can see depends on the position of the Moon in its orbit around the Earth. As a result, the Moon's shape appears to change at different times of the lunar month. These different shapes are called the **phases of the Moon**. When the Moon is between us and the Sun, the whole of the sunlit part of the Moon is facing away from us so we cannot see an illuminated Moon. This phase is called a New Moon. If the Moon is on the opposite side of its orbit, the whole of the sunlit part of the surface is visible from the Earth. This is called a Full Moon.

Fact file for the Moon

- Our Moon is approximately 380 000 km from Earth.

- The Moon has a diameter of 3476 km.

- Gravity on the Moon is one-sixth as strong as that on Earth.

- On the sunlit side of the Moon the temperature is approximately 100 °C. On the dark side it is approximately −170 °C.

- The Moon was formed at the same time as the Earth but it has no atmosphere.

- The Moon orbits the Earth once every 28 days. This is called a **lunar month**.

Between these two positions we see different amounts of the Moon's surface being illuminated, as shown in Figure 9.

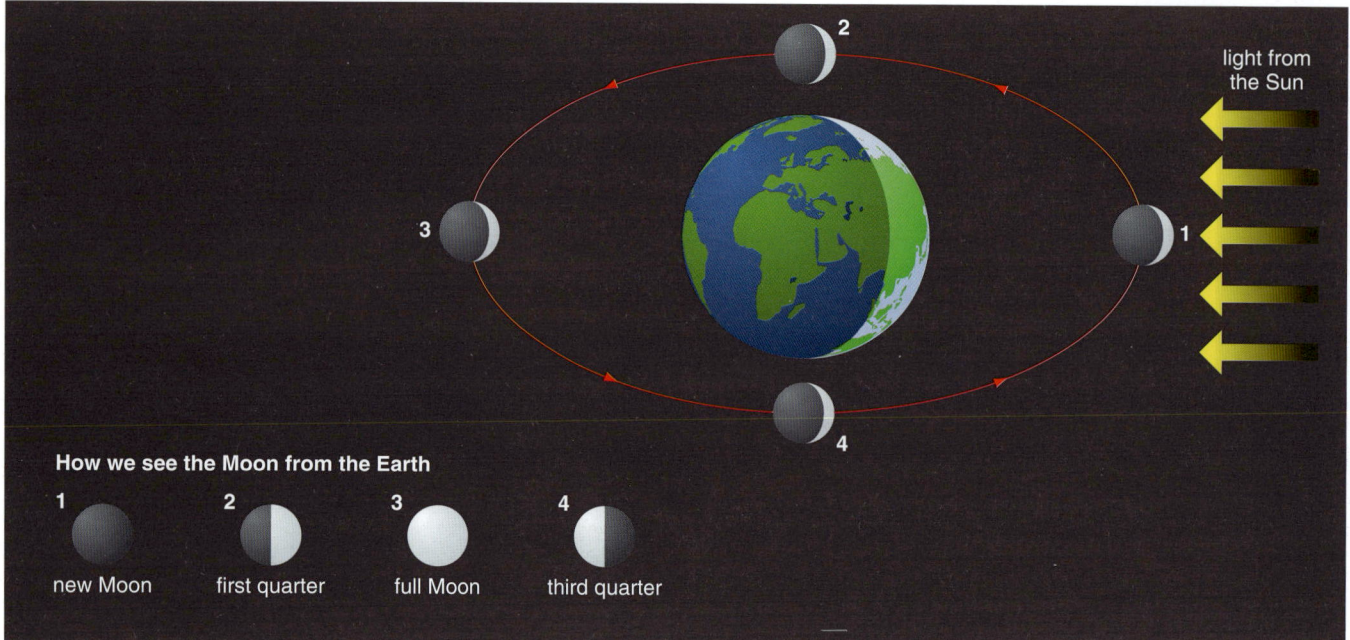

Figure 9 ▲ The different phases of the Moon

Solar and lunar eclipses

If the Moon passes directly between the Sun and the Earth, it blocks off the sunlight and casts a shadow on the Earth's surface. This is called a **solar eclipse**. In those places where all the light is blocked off, a total eclipse is seen. If only part of the Sun's light is blocked off, a partial eclipse is seen.

When the Moon is on the opposite side of the Earth, i.e. furthest from the Sun, it may pass through the shadow cast by the Earth. This is called a **lunar eclipse**.

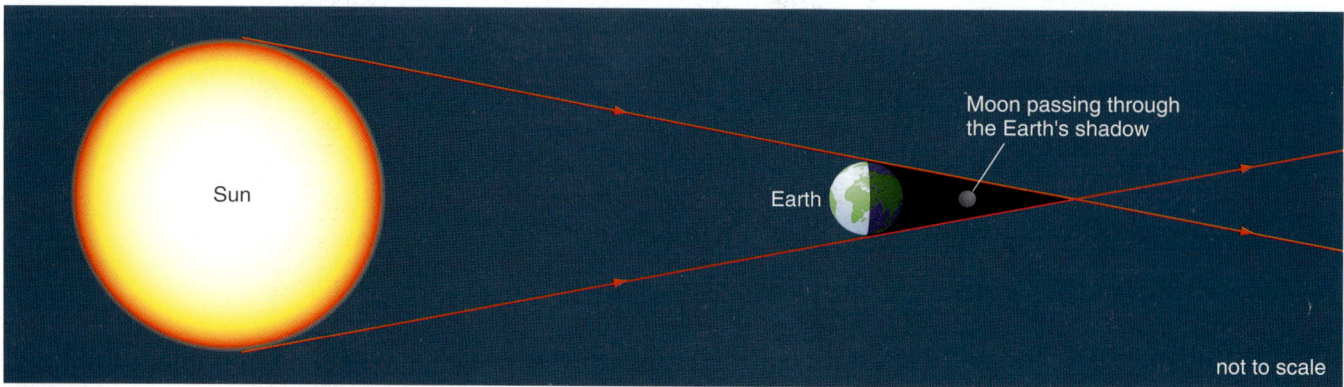

Figure 10 ▲ A lunar eclipse

Tides

The Moon is responsible for the rise and fall in the levels of seas and oceans, i.e. the **tides**. The gravitational pull of the Moon causes the water to bulge on the side nearest the Moon. There is another bulge on the opposite side. As the Earth spins, coastal regions experience high and low tides as the bulges move through them. If the Moon were stationary there would be two high tides and two low tides every 24 hours, but because the Moon is moving around its orbit, this time is 24 hours and 50 minutes. Because this is greater than 24 hours, the time of a high or low tide in a particular coastal region changes each day.

Although the Sun is a lot further away from the Earth than the Moon, it does contribute to the heights of the tides. When the gravitational pull from the Sun and the Moon are in the same direction, we get especially high tides called **spring tides**. When the gravitational forces are at right angles to each other, their effects partly cancel each other out and the tides are lower than normal. These are called **neap tides**.

Figure 11 ▼ The gravitational pull of the Moon is responsible for tides.

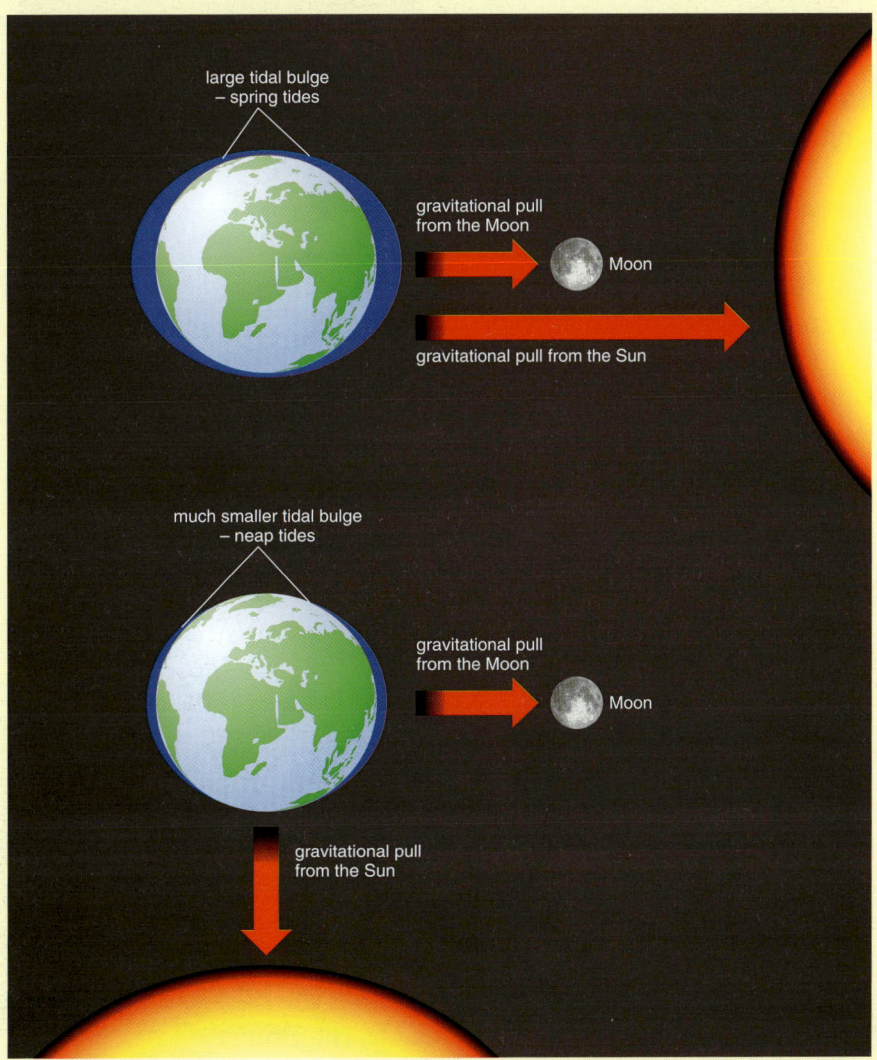

10 What is a moon?

11 Is a moon a luminous or non-luminous object? Explain your answer.

12 What is a Full Moon?

Ideas and Evidence

Landing on the Moon

On 16th July 1969, three American astronauts, Neil Armstrong, Edwin Aldrin and Michael Collins, sat in a small capsule on top of a Saturn rocket waiting for the Apollo 11 mission to begin. Four days later they arrived at their destination and Neil Armstrong became the first man to step onto the Moon. As he did so he uttered those now famous words: 'That's one small step for a man, one giant leap for mankind.' Armstrong and Aldrin walked on the Moon's surface for 2½ hours, collecting samples of dust and rock. They returned safely to Earth on 24th July. There were five further Apollo missions between 1969 and 1972. During this period almost 400 kg of Moon rock was brought back to Earth. On the last three missions, the astronauts used a Moon buggy called the Lunar Rover which allowed them to explore more of the Moon's surface.

Figure 12 ▲ Edwin (Buzz) Aldrin walks on the Moon

Asteroids

Between the planets Mars and Jupiter there is a belt of rock debris. These chunks of rock are called **asteroids** and vary in size from just a few metres across to several hundreds of kilometres across.

Comets

Comets are made of dust and ice. Like planets, they circle the Sun but their orbits tend to take them both very close to the Sun and to the outer edges of the Solar System.

As a comet approaches the Sun, some of its frozen gases evaporate creating a spectacular tail of dust and ice which can range from thousands to millions of kilometres long. Perhaps the most famous comet is Halley's Comet. Its orbit makes it visible from the Earth every 76 years. Halley's Comet was last seen in 1986.

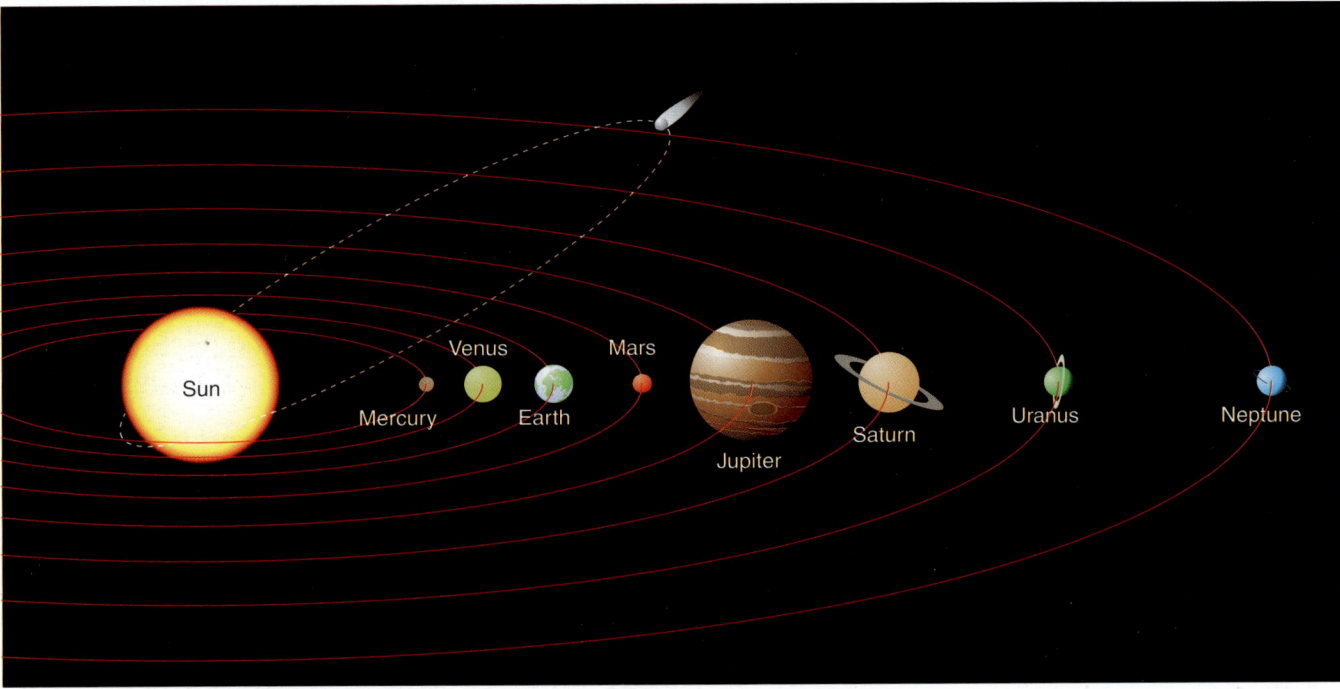

Figure 13 ▲ The orbit of a comet

Figure 14 ◄ The comet Hyakutake, photographed in March 1996

Summary

When you have finished studying this chapter, you should understand that:

✔ The Earth is one of nine planets that orbit a star we call the Sun.

✔ The nine planets in our Solar System, in order of distance from the Sun, are: Mercury, Venus, Earth, Mars, Jupiter, Saturn, Uranus, Neptune, Pluto.

✔ The Earth spins on its axis once every day.

✔ The Sun is much higher in the sky in summer than it is in winter.

✔ It is daytime on the half of the Earth's surface receiving light.

✔ It is night-time on the half that is receiving no sunlight.

✔ The rotation of the Earth makes the Sun appear to rise in the East and set in the West.

✔ The Earth orbits the Sun once every year following an elliptical path.

✔ Because the axis of the Earth is tilted, we experience seasonal changes in temperature and the length of day as the Earth moves around the Sun.

✔ The Sun, planets, asteroids and comets together form our Solar System.

✔ Moons are natural objects which orbit a planet. The Earth has one moon.

✔ If the Moon passes between the Sun and the Earth, we experience a solar eclipse.

End-of-Chapter Questions

1 Explain in your own words the following key terms you have met in this chapter:

planet
day
daytime
night-time
time zone
Greenwich Mean Time
international date line
ellipse
season
leap year
Solar System

Moon
lunar month
non-luminous
phases of the Moon
solar eclipse
lunar eclipse
tides
spring tides
neap tides
asteroids
comets

2 The diagram on page 121 shows the orbits of some of the planets.

a) What shape are these orbits?

b) Name one planet whose year is longer than that of the Earth.

c) What is an asteroid? Where in the Solar System would you find a belt of asteroids?

d) What is a comet? Copy the diagram on page 121 and add to it the path of a comet.

End-of-Chapter Questions continued

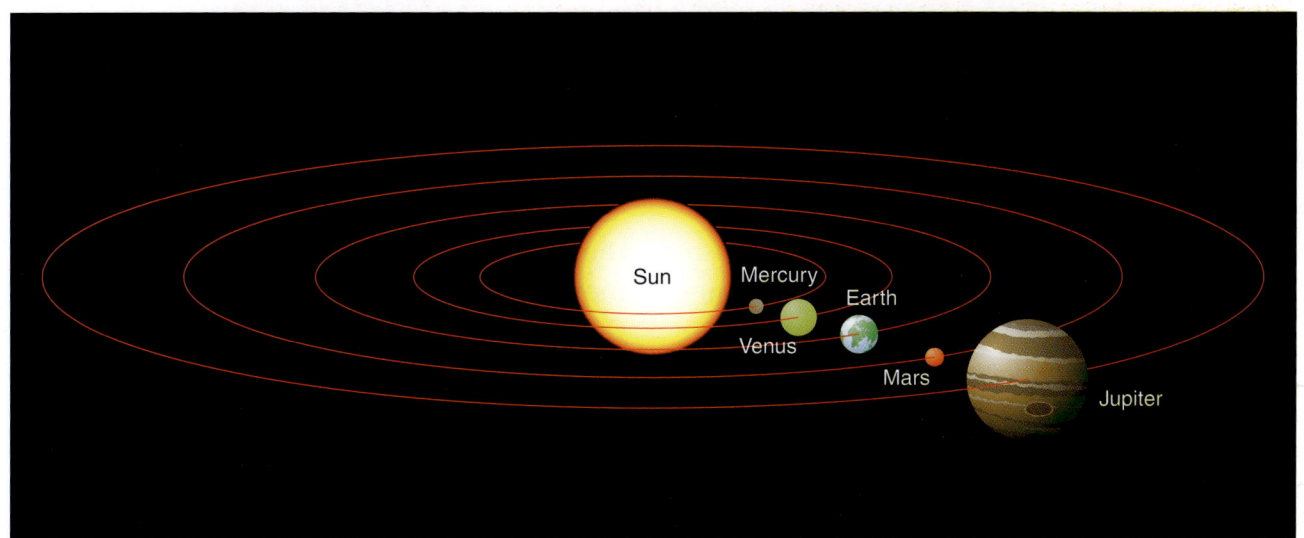

3 The diagram below shows the Earth in four different positions as it orbits the Sun.

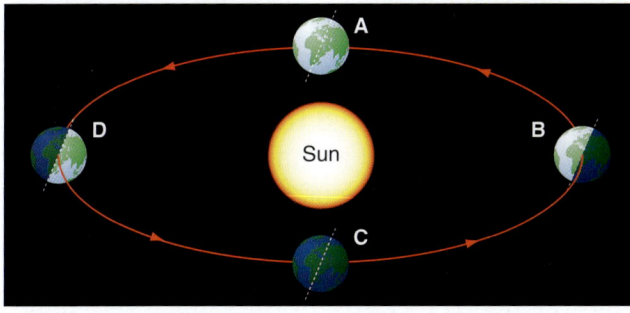

a) How long does it take for the Earth to complete one full orbit of the Sun?

b) In which position is it winter in the Northern hemisphere? Explain your answer.

c) In which position is it autumn in the Southern hemisphere? Explain your answer.

d) What is a leap year? Explain your answer.

4 Draw a diagram of the world, looking down on the North pole.

a) Mark on your diagram the direction in which the Earth rotates.

b) Shade half of your diagram to show that part of the Earth which is in night-time.

c) Now add the following labels to your diagram: (i) X – This is a place where it is midday. (ii) Y – This is a place where it is dawn. (iii) Z – This is a place where it is early evening.

d) Explain why it would have been difficult to create a national rail timetable before the introduction of time zones.

5 Explain in your own words why the times of high tides and low tides at a seaside resort change from day to day.

6 Find out what a meteorite is, then answer the following question: 'Why are there lots of craters visible on the surface of the Moon but there are very few on Earth'.

7 Using the information in Table 1 on page 114, suggest which other planet in the Solar System, besides the Earth, could possibly sustain life. Explain your answer.

The stars and the Universe

For thousands of years, Man has looked up into the skies and wondered at the objects he could see. It was thought by some that the sky was dome-shaped with the stars hanging down from it. Others thought that the stars were distant heavenly bodies.

The Babylonians 3000 years ago believed that our lives were affected by the movements of heavenly bodies. They called this belief **astrology**.

Ideas and Evidence

Models of the Universe

In 120AD, an Egyptian astronomer named Ptolemy suggested that the Earth was at the centre of the Universe and that the Sun, the Moon and all the planets moved around the Earth in circular orbits.

For over 1200 years this model (known as the geocentric model) went unchallenged until a Polish astronomer named Nicolas Copernicus dared to suggest that the Sun was at the centre of our Solar System and that the Earth and the other planets moved around it in circular orbits. When Copernicus first suggested his theory, church leaders were furious because this challenged their belief that God had placed the Earth at the centre of the Universe. Their opposition to Copernicus' model lead to his book being banned and his ideas being discarded.

Nearly 100 years later, in the 17th Century, telescopes became more readily available and the observations made by astronomers of the time confirmed the Sun-centred, or heliocentric, model.

Johannes Kepler, a German astronomer of the time, took this model one step further when, through careful observation of the motion of the planets, he came to the conclusion that a planet's orbit is **elliptical** rather than circular. An elliptical orbit looks like a slightly squashed circle.

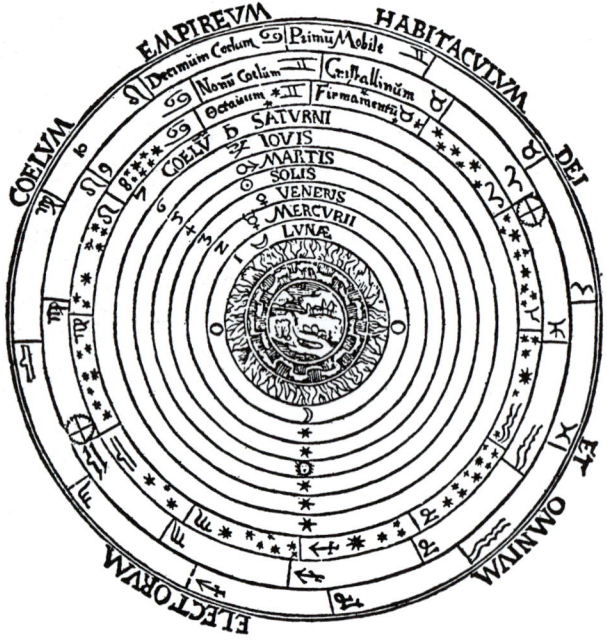

Figure 1 ▲ Ptolemy's geocentric view of the Universe with the Earth at the centre

Ideas and Evidence

Models of the Universe Continued

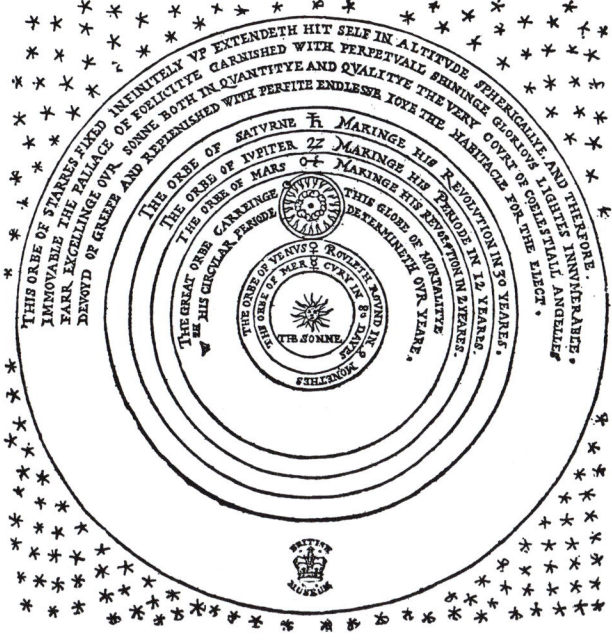

Test Yourself

1 Who first suggested that the Earth was at the centre of the Universe?

2 Who first suggested that the Sun was at the centre of the Solar System?

3 What change to the previous models of the Solar System did Johannes Kepler make?

Figure 2 ◄ Copernicus' heliocentric view of the Universe with the Sun at the centre

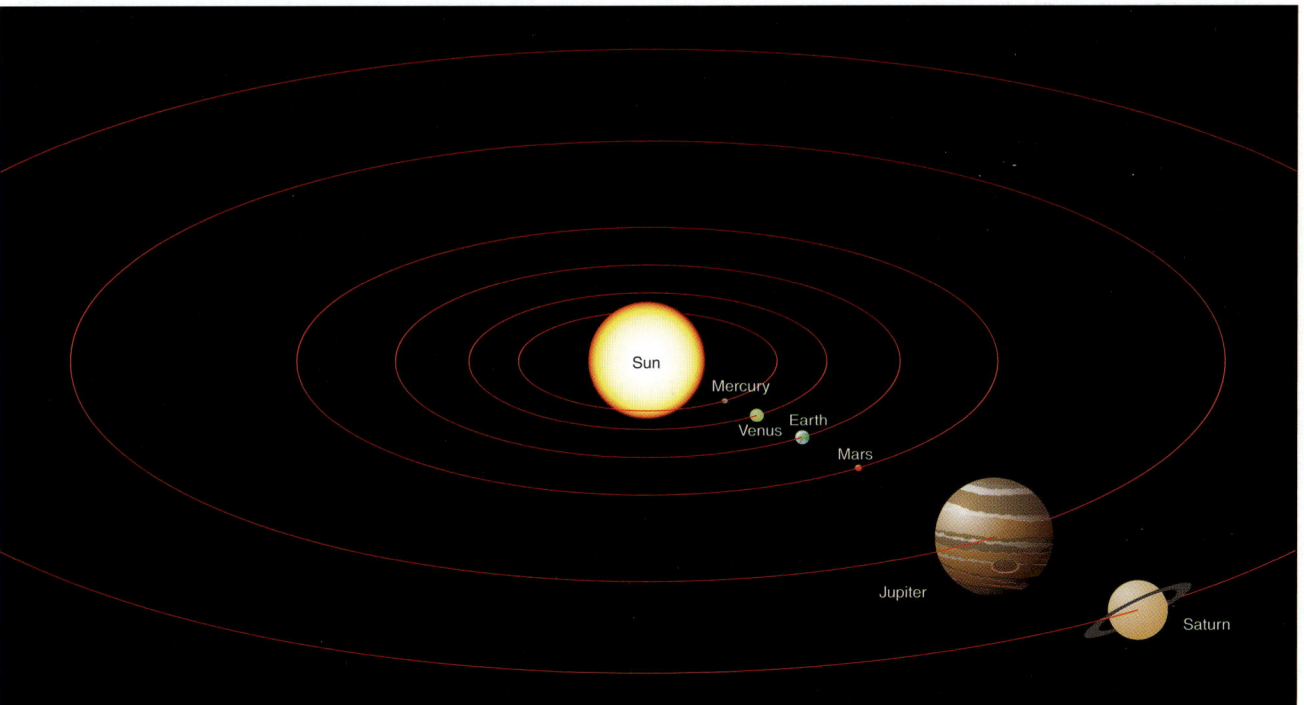

Figure 3 ▲ Kepler's model of the Solar System had the planets moving in elliptical orbits around the Sun

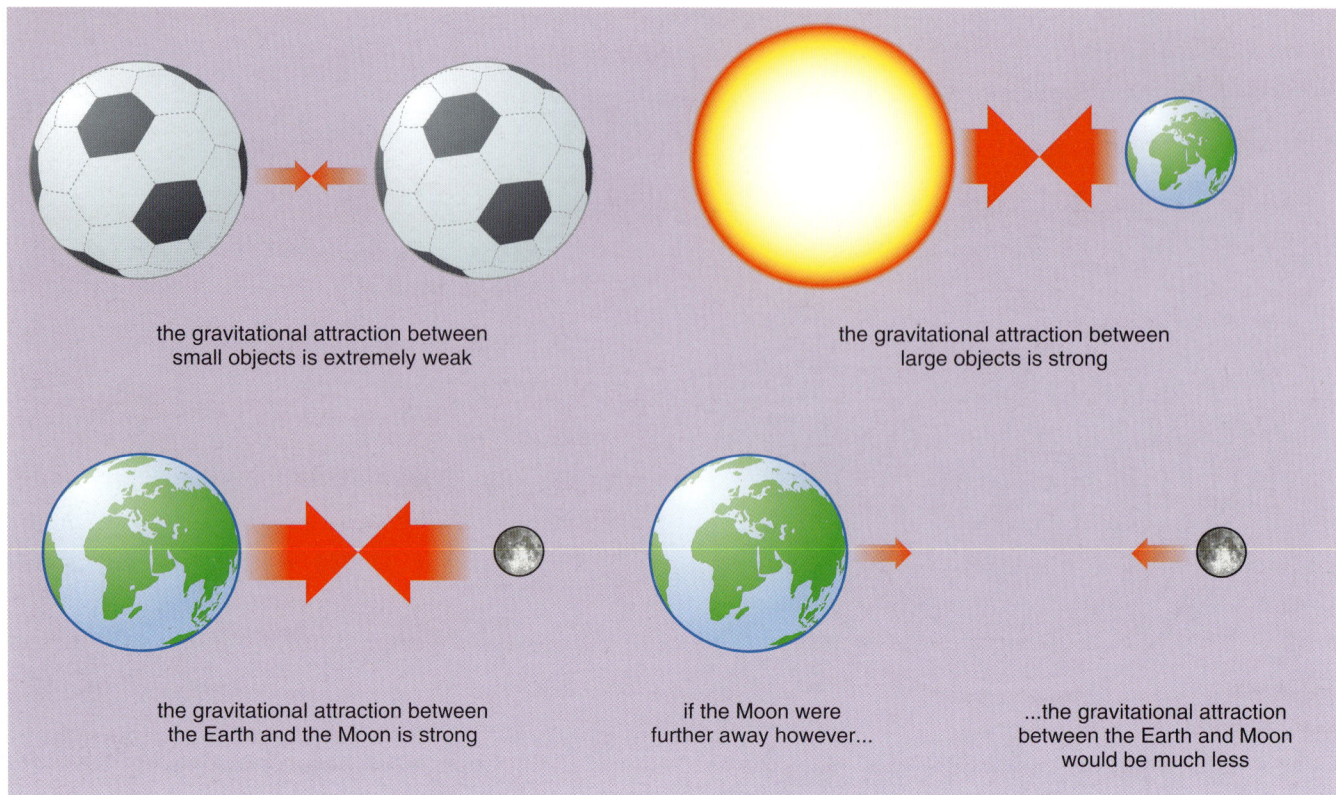

the gravitational attraction between
small objects is extremely weak

the gravitational attraction between
large objects is strong

the gravitational attraction between
the Earth and the Moon is strong

if the Moon were
further away however...

...the gravitational attraction
between the Earth and Moon
would be much less

Figure 4 ▲

Gravity

In 1687, a scientist called Isaac Newton put forward a theory
which not only explained the motions of all the bodies in the sky,
but eventually explained how the stars and their planets were
first formed.

Newton suggested that all objects are attracted to all other
objects by forces called **gravitational forces**. The size of these
forces depends upon the masses of the objects and their
separation. The larger the mass of the objects, the stronger the
forces. The closer the objects, the stronger the forces.

Moving along a curved path

If an object is to move along a curved path, for example a circle, a
force must be applied to it.

Planets travel along curved paths around the Sun. There must,
therefore, be attractive pulling forces between the Sun and the
orbiting planets. Our Sun contains 99% of the mass of our Solar
System. It is very large. According to Newton it will, therefore,
apply large attractive forces to any object that is near by.

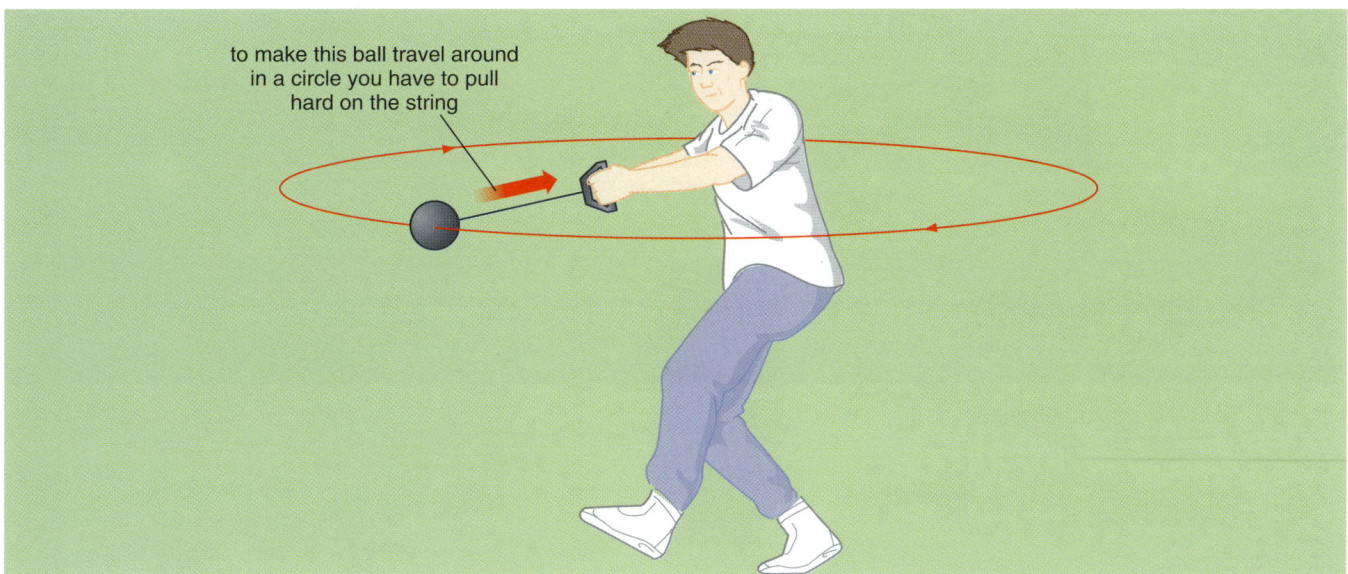
to make this ball travel around in a circle you have to pull hard on the string

Figure 5 ▲ If this ball is to travel round in a circle, the athlete must apply a force to it. He must pull on the chain as he spins the ball around. If he stops applying a force, i.e. he lets go, the ball ceases to go around the circle and moves away along a straight line

It is these gravitational forces that cause the planets to move in orbits and hold our Solar System together. Planets close to the Sun experience large forces and so follow very curved paths. Planets further from the Sun experience weaker gravitational forces and so follow less curved paths. Moons orbit planets. They are held in these orbits by the gravitational attraction between them and their planet.

Test Yourself

4 Name one planet in our Solar System which experiences large gravitational forces from the Sun.

5 Which planet in our Solar System experiences the weakest gravitational forces from the Sun?

6 What shape of path would a planet follow if there were no gravitational forces?

Stars

Stars are **luminous** objects – they give out their own light. Our nearest star is the Sun. It is a very average star. Its surface temperature is approximately 6000 °C, whilst the temperature at its centre is about 15 million °C. These high temperatures are the result of nuclear reactions.

Stars like our Sun are formed from giant clouds of dust and gas. These particles are gradually drawn together over many millions of years by gravitational forces. As they come together, nuclear reactions begin and a very hot ball of gas forms which we call a star.

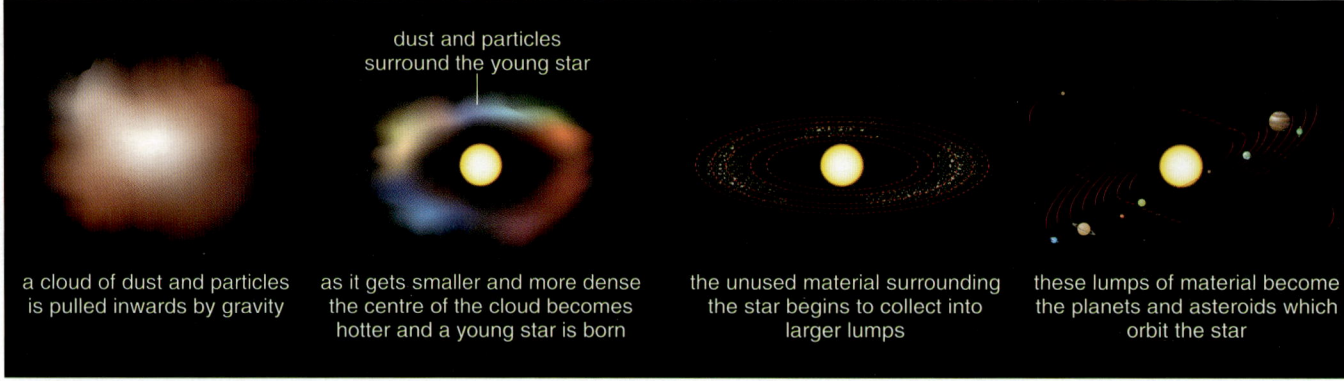

dust and particles surround the young star

| a cloud of dust and particles is pulled inwards by gravity | as it gets smaller and more dense the centre of the cloud becomes hotter and a young star is born | the unused material surrounding the star begins to collect into larger lumps | these lumps of material become the planets and asteroids which orbit the star |

Figure 6 ▲   The birth of a star

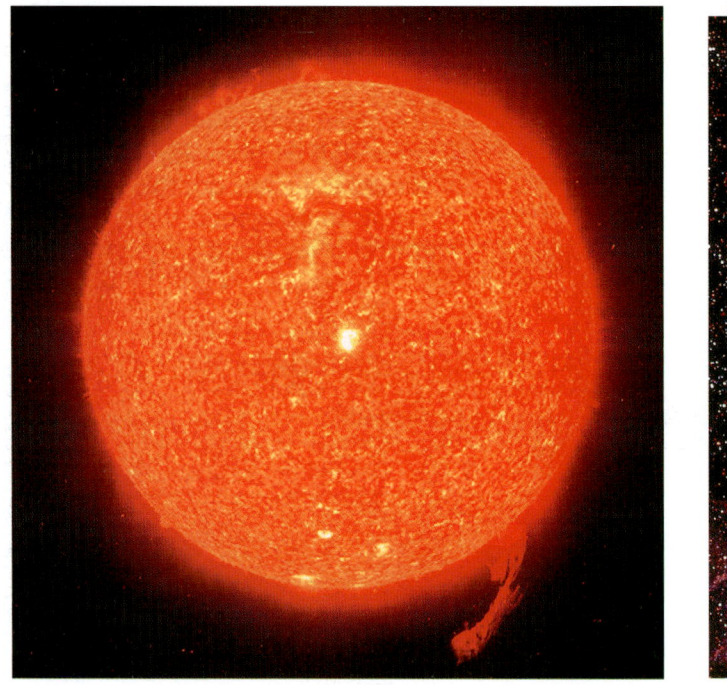

Figure 7 ▲   Our Sun

Figure 8 ▲   A supernova

Some of the dust and gas particles may come together to form smaller bodies which then orbit the star. We call these bodies planets.

Stars are not permanent features in our skies. They are gradually changing but these changes can take billions of years. Some stars end their lives 'quietly', gradually cooling to form stars that do not emit any light. Other stars end their lives with an enormous, spectacular explosion called a **supernova**. A supernova can be as much as 500 million times brighter than the Sun.

Fact file for the Sun

- The diameter of the Sun is 1 392 000 km.

- Our Earth could be squashed inside the Sun a million times.

- Sunspots are regions of the Sun's surface that are cooler than the rest.

- The number of sunspots varies with time reaching a maximum every 11 years.

- Eruptions of gas called prominences continuously rise from the Sun's surface. Some are as much as 2 million km long.

- The Sun has provided the Earth with all its energy needs for thousands of millions of years.

- Four million tonnes of the Sun's material is turned into energy every second.

Test Yourself

7 From what materials are stars formed?

8 What happens to these materials?

9 What kinds of reactions produce the energy which is emitted by a star?

10 What is a sun spot?

Constellations and galaxies

During the daytime, the light from the Sun is so bright it is impossible to see any other stars. They are only visible after the Sun has set. Even then these other stars seem small and dim compared with our Sun. This is because they are much further away.

Approximately 1500 stars are visible to the naked eye in the night sky. There are groups of stars which may appear to be close together. These are called **constellations**. You will probably have seen and heard of some of them. Astronomers often join up the stars with imaginary lines to make their shapes easier to see.

Stars and constellations cluster together in enormous groups called **galaxies**. The galaxy we live in is called the **Milky Way**. It contains about 200 000 million stars. It is now accepted that there are billions of galaxies spread throughout the Universe.

Figure 9 ▲ Some well-known constellations

Figure 10 ▶ The Milky Way

Test Yourself

11 Name three constellations.

12 Explain the difference between a constellation and a galaxy.

13 What is the name of the galaxy in which we live?

Figure 11 ▲ This time lapse photo of the night sky shows how the stars appear to move in circles

The moving night sky

If we were to take a photograph of the stars in the sky every hour, we would see that the stars appear to change their position, moving in circles. This happens because the Earth is rotating. Only one star, the Pole Star, appears not to move as it is directly above the Earth's axis of rotation.

Because the Pole Star does not change its position it was used by ancient travellers to help them navigate.

Light years

Distances between objects in space are so large that our normal unit of measurement, the kilometre, is far too small. We need to use a unit of distance that is much larger. The Sun is 150 million kilometres from the Earth. Light from the Sun takes just 8.3 minutes to travel this distance. We use this idea of how long it will take light to travel between two points as our new measure of distance. It is called a **light year**. A light year is the distance a ray of light will travel in one year.

Our second nearest star (after the Sun) is so far away, light takes 4.3 years to reach the Earth. We say it is 4.3 light years from Earth. Light from our nearest galaxy, the Andromeda galaxy, takes 2 million years to reach us.

Object	Distance from Earth in light years
Sun	8.3 light minutes
Proxima Centauri (our second nearest star)	4.3 light years
Rigel	815 light years
Andromeda galaxy	2 million light years

Table 1 ▲

Exploring space

Although Man has shown by his visits to the Moon and the building of space stations that orbit the Earth that he is able to travel in space, **artificial satellites** and probes offer a cheaper and safer way of investigating our Solar System and beyond.

Artificial satellites

A **satellite** is an object which orbits a planet. Here on Earth we have one natural satellite, the Moon. An artificial satellite is a man-made object that orbits the Earth. Artificial satellites have many uses:

- to observe the Earth from above e.g. weather satellites.

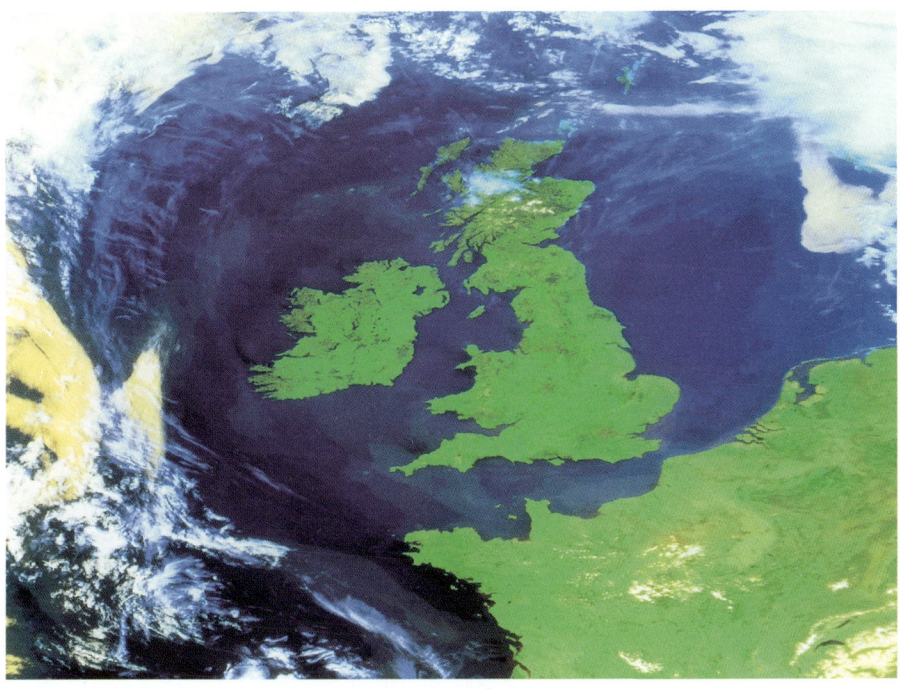

Figure 12 ◄ This photograph was taken by a weather satellite orbiting the Earth

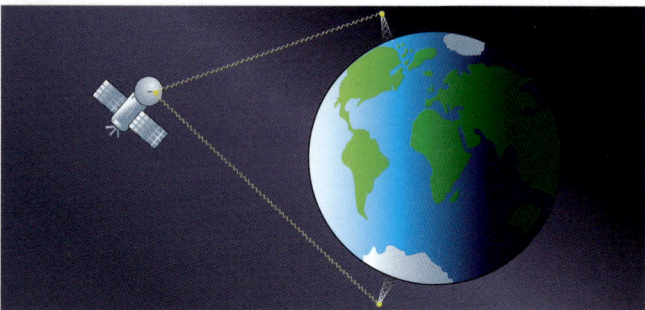

Figure 13 ▲ When the radio signal arrives at the satellite, it is redirected down to another part of the Earth where the message is received

Figure 14 ▶ Above the Earth's atmosphere, the Hubble telescope is able to take photographs of previously unseen objects, like distant galaxies

- to look away from the Earth into space without having to look through the Earth's atmosphere e.g. the Hubble telescope.

- to allow communication between people all over the world using radio, television or the Internet.

Table 2 on page 131 lists some of the main space probes that have been used to explore space.

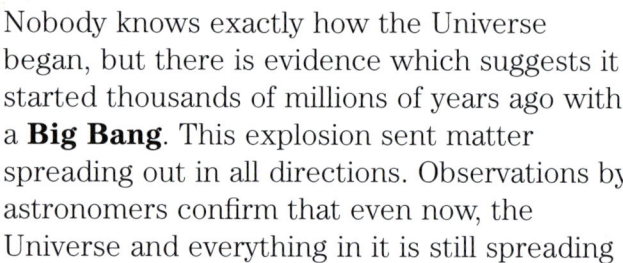

Ideas and Evidence

Origins and future of the Universe

Nobody knows exactly how the Universe began, but there is evidence which suggests it started thousands of millions of years ago with a **Big Bang**. This explosion sent matter spreading out in all directions. Observations by astronomers confirm that even now, the Universe and everything in it is still spreading out. Some scientists believe that the Universe will continue to expand forever. This theory is called the continuous expanding universe. Others believe that the forces of gravity will slow this expansion down and eventually pull all the matter back together.

Test Yourself

14 Which is the only star which appears not to move in the night sky? Why does this star appear to be in a fixed position?

15 For what purpose was this star used by ancient travellers?

16 Why do we not use kilometres to measure the distance between stars? What do we use instead?

Year	Space probe
1959	Lunar 1, 2 and 3 were the first probes to take close-up photographs of the Moon
1965	Mariner 4 took photographs of Mars
1966	Lunar 9 landed on the Moon
1967	Venera 4 landed on Venus
1971	Mariner 9 was put in orbit around Mars and took detailed photographs of its surface
1976	Vikings 1 and 2 landed on Mars
1978	Venera 11 and 12 and two pioneer probes visited Venus
1979	Voyager 1 passed close to the planet Jupiter
1980	Voyager 1 passed close to the planet Saturn
1986	Voyager 2 passed close to the planet Uranus
1986	Four probes were launched to investigate Halley's Comet
1989	Voyager 2 passed close to Neptune
1989	Phobos 2 entered orbit around Mars Galileo was launched to visit the planet Jupiter
1991	Magellan sent back detailed information about the surface of Venus
1999	Voyager 1 becomes the furthest man-made object in space. It has been travelling for approximately 22 years
2001	Galileo passes within 200 km of Io, one of Jupiter's moons

Table 2 ▲

Year	Event
1957	Launch of Sputnik 1, the first object to orbit the Earth
1957	First animal in space, a dog called Laika
1960	First animals to return from space flight, two Soviet dogs
1961	First person to orbit the Earth, Yuri Gagarin
1963	First woman in space, Valentina Tereshkova
1969	First space walk
1969	First man on the Moon, Neil Armstrong
1971	First space station launched
1981	First space shuttle flight, Columbia
1986	The Soviet space station Mir is launched
1990	The Hubble Space Telescope is put into orbit
2001	Dennis Tito becomes the first 'space tourist'

Table 3 ▲   Some important dates in the history of space travel

Summary

When you have finished studying this chapter, you should understand that:

✔ Our ideas about the Universe have changed as scientists have been able to make more accurate observations and measurements of the heavenly bodies.

✔ Scientists use artificial satellites and space probes to improve our knowledge of the Universe.

✔ Stars and planets begin their lives as particles of dust and gas that are pulled together by gravitational forces.

✔ Stars are luminous objects. Our nearest star is the Sun.

✔ Stars are not normally visible during the day because the Sun is very bright. Some groups of stars appear close together. We call them constellations.

✔ Stars appear to move across the night sky because of the rotation of the Earth.

✔ Distances in space are very large so we measure them in light years. A light year is the distance a ray of light will travel in one year.

✔ Scientists believe that the Universe began with a large explosion, the Big Bang.

End-of-Chapter Questions

1 Explain in your own words the following key terms you have met in this chapter:

astronomy   galaxy
ellipse   Milky Way
gravitational forces   light year
star   artificial satellite
luminous   satellite
supernova   Big Bang
constellation

2 Starting with the smallest, put these objects in order according to their size:

galaxy, universe, planet, constellation, star, comet

Name two objects in the above list that are luminous.

3 Draw a diagram to show how a communication satellite can be used so that radio signals sent from one side of the Earth can be received on the other side. Why is the Hubble telescope able to take better photographs of objects in space than telescopes that are on the Earth's surface?

4 The diagram below shows the path of a comet.

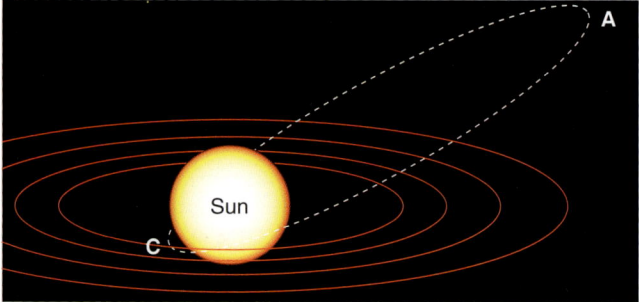

a) In which position will the gravitational forces on the comet due to the Sun be greatest?

b) Give a reason for your answer to part a).

c) What happens to the speed of the comet as it moves i) from position A to position C and ii) from position C to position A?

d) Describe one difference between the paths of planets and the paths of comets.

5 What is a light year? Give two reasons why unmanned probes are used to explore space rather than manned flights.

6 Find out the difference between astrology and astronomy.

13 Pressure

Sometimes the result of applying a force to an object depends not only on the size of the force but also on how it is distributed. When the coin is held vertically and pushed downwards, the force being applied to the plasticine through the coin is concentrated on the small surface area of the edge of the coin. A large **pressure** is created here and the coin sinks into the plasticine. When the coin is flat, the applied force is spread over a larger area and so creates a smaller pressure. It is much more difficult now to push the coin into the plasticine

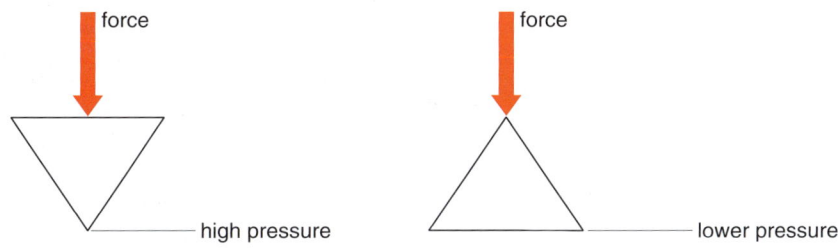

Figure 1 ◄ If a force is concentrated into a small area, it creates a large pressure. If a force is spread out over a large area, it creates a smaller pressure

The force applied by the woman in Figure 2 is more concentrated i.e. a higher pressure is created under her stiletto heels and so this is much more painful.

Calculating pressures

We calculate the size of the pressure created by a force using the equation

$$\text{Pressure (P)} = \frac{\text{Force (F)}}{\text{Area (A)}}$$

If the force is in newtons and the area is in m^2, the pressure is in N/m^2 or **pascals** (Pa).

If the force is in newtons and the area is in cm^2, the pressure is in N/cm^2.

Figure 2 ▲ If a man wearing normal shoes were to tread on your foot it would hurt, but not as much as if a woman of the same weight wearing stilettos stood on your foot!

Example

Calculate the pressures created by a coin when a force of 10 N is applied to a) its flat surface (A = 2.5 cm^2) and b) its edge (A = 0.5 cm^2).

a) P = F/A
    P = 10 N/2.5 cm^2
    P = 4.0 N/cm^2

b) P = F/A
    P = 10 N/0.5 cm^2
    P = 20 N/cm^2

Test Yourself

1   When do the forces applied to an object produce a) high pressure and b) low pressure?

2   A man weighing 1000 N stands on a piece of wood on the ground. The wood is square and measures 2 m × 2 m. Calculate the pressure created on the ground by the man's weight.

Creating high pressures

Sometimes we want to create high pressures . . .

- A cheese cutter is able to slice easily through cheese because the forces applied to it are concentrated on the very small surface area of the wire and so create very large pressures.

- Walking or climbing in icy conditions can be difficult. It is much safer if you wear crampons. Your weight is then concentrated on the very small surface area of the spikes. Very large pressures are therefore created here which allow the spikes to dig into the hard surface of the ice, giving you better grip.

- The point of a drawing pin has a very small surface area. A force applied here will create a pressure large enough for the point to pierce the surface. The head of a drawing pin has quite a large surface area. The pressure here is, therefore, quite small and not painful to the thumb.

Keeping the pressure down

At other times we want to create low pressures . . .

- Wearing snowshoes or skis increases the surface area over which your weight is spread. Because the pressure beneath you is less, you are unlikely to sink into the snow.

Figure 3 ▲   This climber is wearing crampons which help him to grip the ice

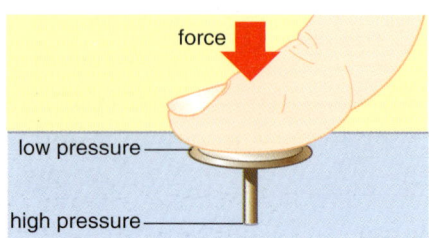

Figure 4 ▲

Figure 6 ▲   This rescuer is using a plank to reduce the pressure he exerts on the ice

Figure 5 ▲   Snowshoes spread out your weight so you can safely walk on the snow

- Camels have adapted in several ways to help them survive in desert conditions. Having large feet helps them walk over sandy terrain without sinking into it.

- You should never walk across a frozen pond or lake. The pressure your weight creates may be sufficient to crack the ice. Rescuers can avoid the problem by using a long ladder or plank of wood which spreads their weight and reduces the pressure they exert on the ice.

- If a tractor is working on boggy ground, it uses very wide tyres to prevent it from sinking into the mud.

- Because there are a large number of nails on this fakir's bed, the pressure on each nail is too small to pierce his skin.

Figure 7 ▲   Camels have large feet to prevent them sinking into the sand

Figure 8 ◄

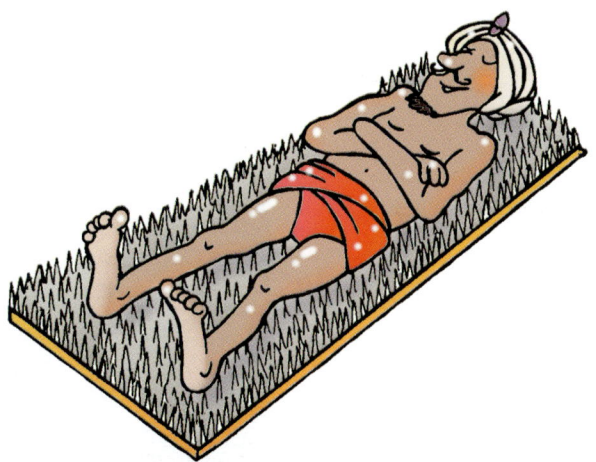

- Carrying heavy objects in carrier bags that have narrow handles can be painful. To avoid these high pressures, bags should have wide handles.

3 Explain why a wire cheese cutter is able to cut through a piece of cheese so easily.

4 Why do rescuers use ladders to reach people who have fallen through ice on frozen lakes and ponds?

5 Explain how and why camels have adapted to allow them to walk more easily in desert conditions.

Figure 9 ▲

Pressure in liquids

The diver in Figure 9 feels **water pressure** all around him. His diving suit will help him survive the high pressures he will feel deep under the surface of the water. The deeper he dives, the greater the water pressure around him.

The pressure the diver feels is the same in all directions. We can demonstrate this with the experiment shown in Figure 10.

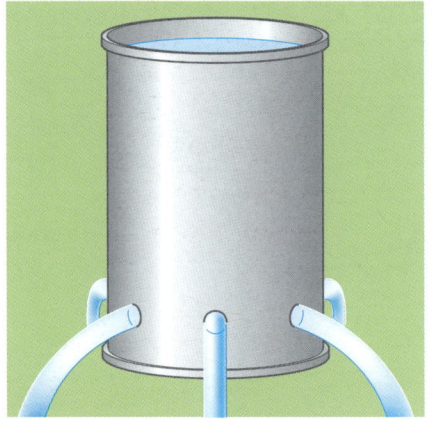

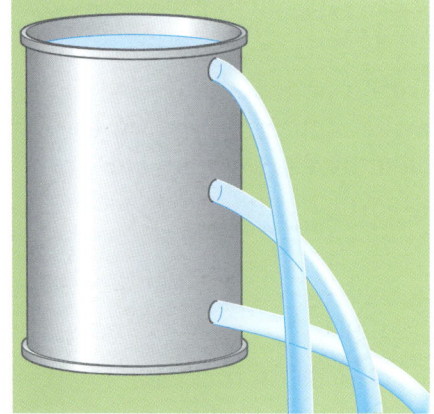

Figure 10 ▲ Water flows out of this can at the same rate in all directions, showing that the pressure in a liquid is the same in all directions

Figure 11 ▲ Water flows out of this can at a faster rate the lower the hole, showing that the pressure in a liquid increases with depth

A similar experiment (Figure 11) clearly shows that the pressure in a liquid increases with depth.

Figure 12 is a photograph of a deep sea submersible. It is used to investigate those parts of our oceans that are several kilometres deep. It was, therefore, designed to withstand extremely high pressures. At a depth of 11 000 m it must withstand pressures of 15 000 psi (pounds per square inch).

Dams have to withstand the very large pressures created by the water they are holding back. Dams are always much thicker at their base because this is where the pressures are greatest.

Figure 12 ▲

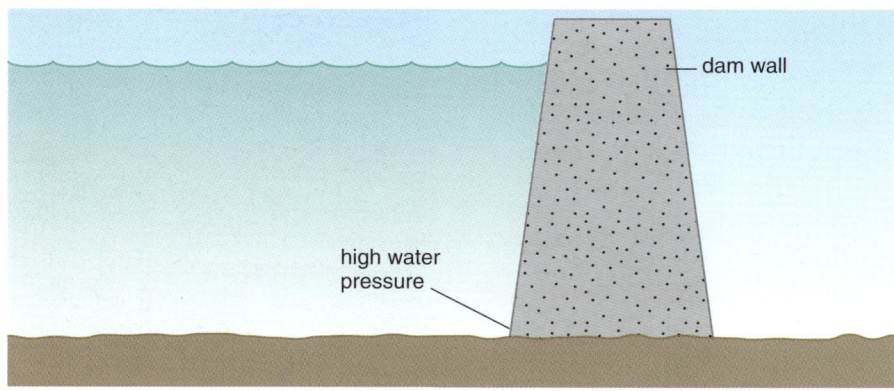

Figure 13 ▲　A cross-section of a dam

Transmitting pressure and forces through liquids

Liquids are **incompressible**. Their particles cannot easily be squashed closer together. As a result, if a force is applied to a liquid, it is **transmitted** or passed through the liquid. The study of transmitting forces and pressures through liquids is called **hydraulics**. This phenomenon can be extremely useful as the examples below show.

Imagine applying a force to each of these systems by pushing in the piston P. The transmitted forces will cause the other pistons to be pushed outwards.

The hydraulic jack

When the handle is pushed down, it applies a force on the small cylinder which creates a pressure in the liquid. This pressure is transmitted to a bigger second cylinder where a much larger force is applied to the second piston i.e. a small force has been changed into a much larger one. The **hydraulic jack** is a **'force multiplier'**.

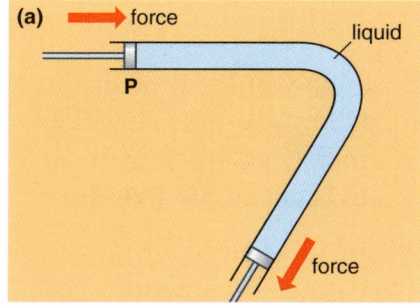

a) The direction of a force is easily changed

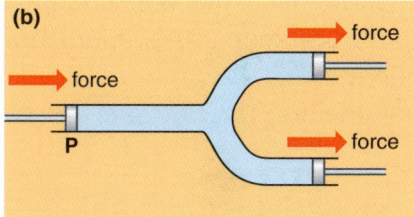

b) The same force can be applied in several different places at the same time e.g. on all four wheels of a car

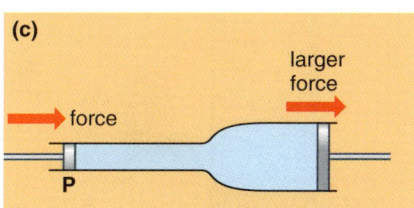

c) The size of the force can be changed.

Figure 14 ▲

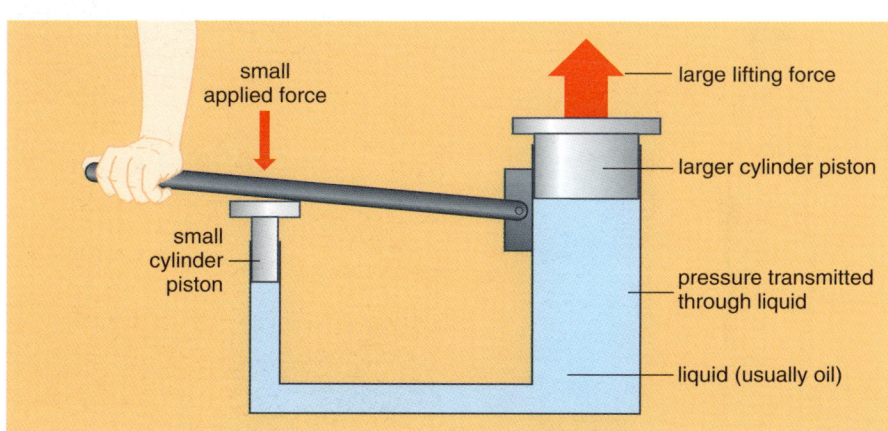

Figure 15 ▲　A hydraulic jack

Hydraulic brakes

The **hydraulic brakes** on a car work in a very similar way. When the driver presses the brake pedal, his small force is increased and directed to all those places where it is needed i.e. to the four wheels.

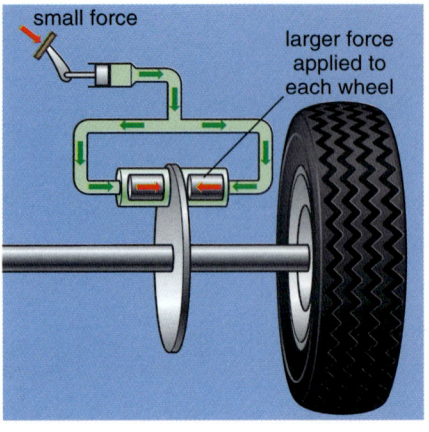

Figure 16 ▲   The hydraulic brake system of a car

Test Yourself

9 Why are liquids incompressible?

10 Give three advantages of transmitting forces through liquids.

11 Why is a hydraulic jack known as a force multiplier?

Pressure in gases

Although you are probably unaware of it, you are having pressure applied to you right now. The air particles around all of us are continually colliding with our bodies. The pressure these particles create is called **air pressure**.

We are so used to air pressure, we are rarely aware it is there but it is easy to show that it exists.

When the top of the bottle in Figure 17 is removed, there is air inside and outside its walls. The pressures created by the colliding air molecules are equal and balanced. The bottle therefore keeps its shape. If some or all of the air inside the bottle is removed, the pressure inside is now less than the air pressure outside and the bottle is crushed.

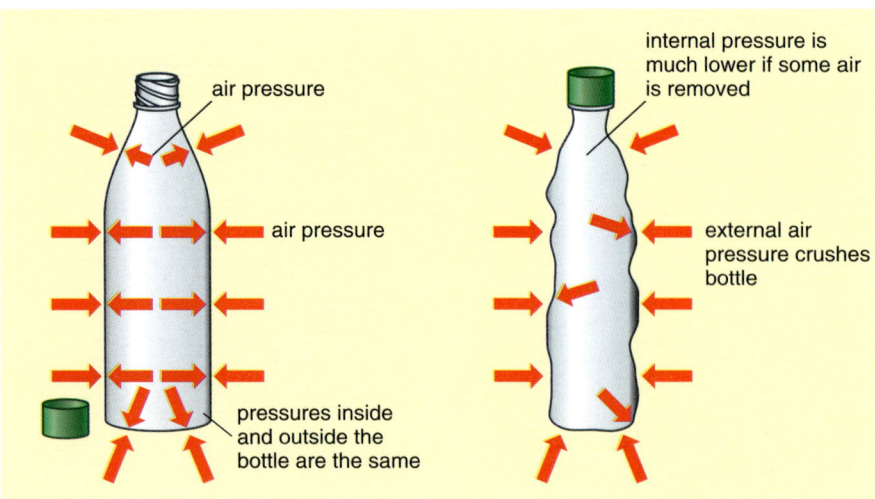

Figure 17 ▲   Demonstrating air pressure

Ideas and Evidence

Atmospheric pressure is quite large. This was demonstrated by Otto von Guericke in 1651. He made two hollow copper hemispheres which, when joined together, formed an airtight sphere with a diameter of about 30 cm. One of the hemispheres had a tap which could be connected to a vacuum pump. With the tap open, the air was pumped out. The tap was then closed and the vacuum pump removed. The two hemispheres were now being held together by just atmospheric pressure. Using two teams of eight horses, von Guericke was unable to pull the hemispheres apart!

Drinking through a straw

How does atmospheric pressure help you drink using a straw?

When you initially suck through a straw, you remove some of the air particles inside it. The pressure exerted by the air particles on the surface of the liquid outside the straw is larger. So liquid is pushed up the straw.

Rubber sucker

How does atmospheric pressure make a rubber sucker stick?

Test Yourself

12 Explain in your own words what causes atmospheric pressure.

13 Describe a simple experiment to demonstrate that atmospheric pressure exists.

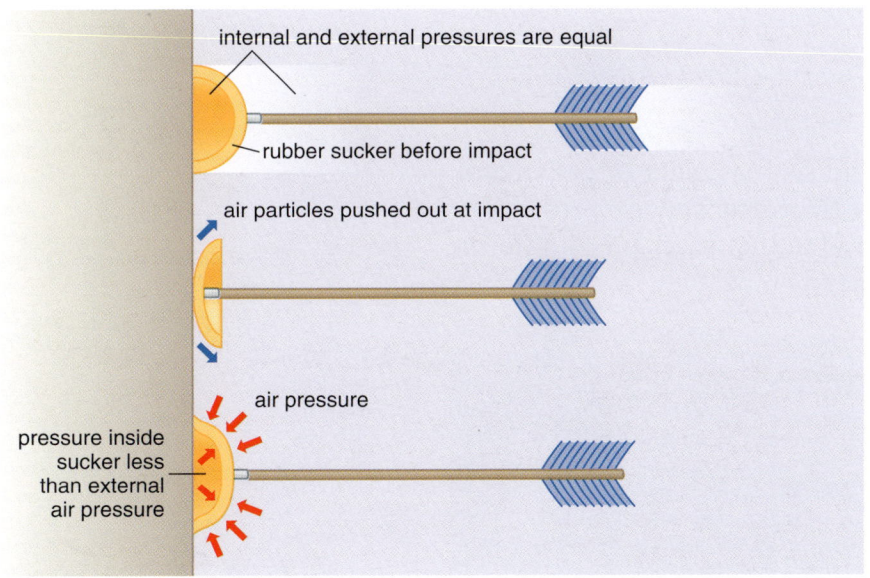

internal and external pressures are equal

rubber sucker before impact

air particles pushed out at impact

air pressure

pressure inside sucker less than external air pressure

Figure 19 ▲

Before striking the wall, the air pressure inside and outside the sucker is the same. As the sucker strikes the wall, it flattens and air particles are pushed out. After the collision, the sucker resumes its initial shape but the pressure inside is now much less than that outside. It is this pressure difference that holds the sucker in place.

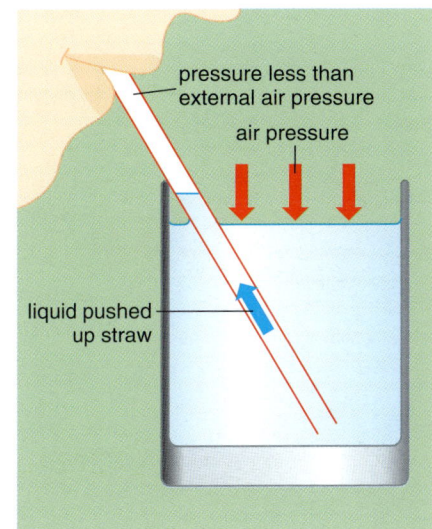

pressure less than external air pressure

air pressure

liquid pushed up straw

Figure 18 ▲

Breathing

How does air pressure help you to breathe?

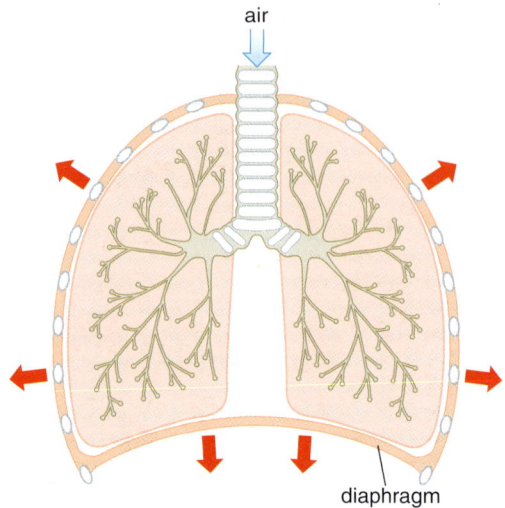

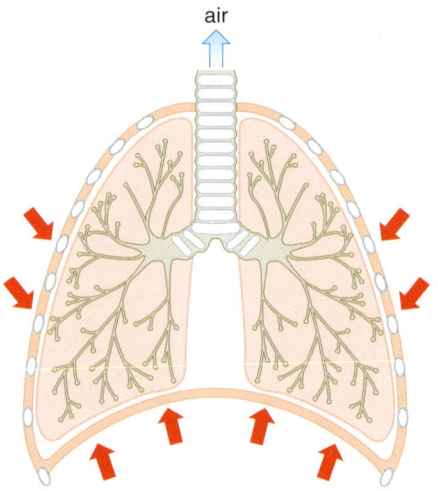

Figure 20 ▲　When you breathe in, your ribcage moves upwards and outwards and your diaphragm moves down, decreasing the pressure in your chest cavity. Air pressure outside your body is higher, so air enters your lungs, causing them to inflate

Figure 21 ▲　When you breathe out, your ribcage moves downwards and inwards and your diaphragm moves up, increasing the pressure in your chest cavity. The pressure here is now greater than the air pressure outside your body, so your lungs are squashed, pushing the air out

Aerosols

An aerosol spray can contains a liquid such as paint or polish and a gas, which is at a pressure greater than atmospheric pressure. When the nozzle is pressed, the liquid in the can is pushed up the tube by the gas and out of the nozzle opening.

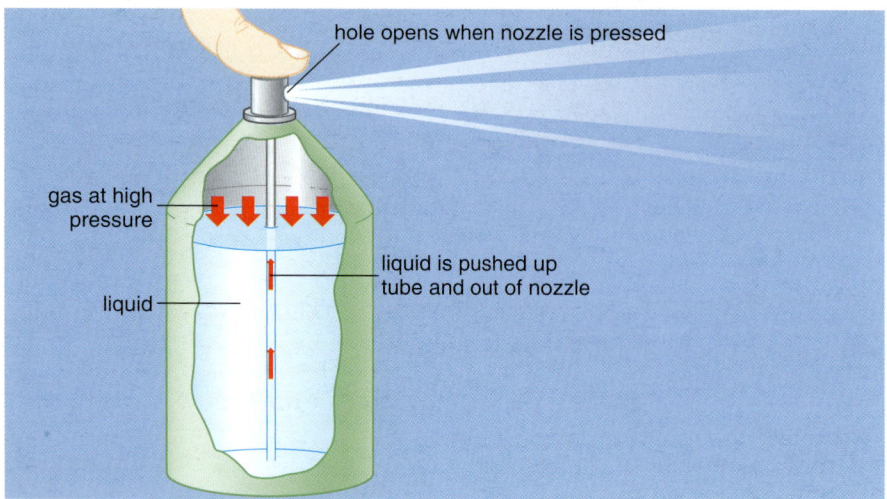

Figure 22 ▲　How an aerosol works

Extension box

Barometers

We measure atmospheric pressure using a barometer. There are two main types of **barometer**. These are mercury barometers and aneroid barometers.

A mercury barometer consists of a tube filled with mercury inverted over a bowl that also contains mercury. The height of the column of mercury in the tube indicates the atmospheric pressure. If atmospheric pressure decreases, the height of the column decreases. If atmospheric pressure increases, the height of the column increases. Standard atmospheric pressure is the equivalent of a column of mercury 760 mm high.

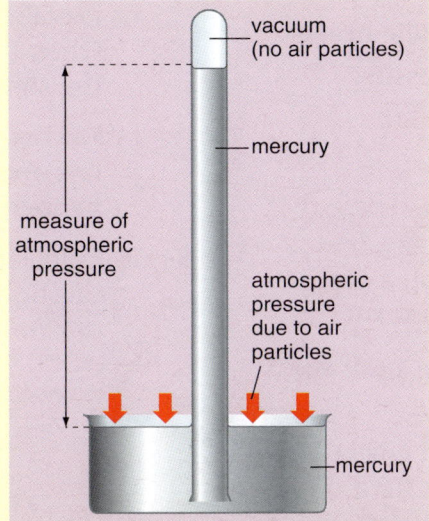

Figure 23 ▲ A mercury barometer

An aneroid barometer contains a partially evacuated box which is squashed if atmospheric pressure increases and expands if it decreases. These small changes in volume are magnified by levers, which then use a chain to move a pointer over a scale.

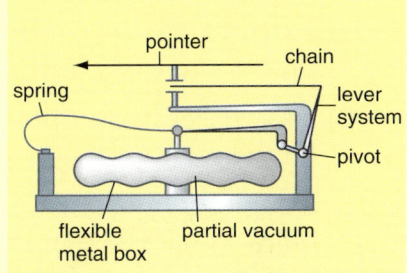

Figure 24 ▲ An aneroid barometer

Test Yourself

14 Explain in your own words how you are able to drink a liquid using a straw.

15 What instrument would you use to measure atmospheric pressure?

Summary

When you have finished studying this chapter, you should understand that:

✔ The effect of a force may depend on the area over which it is applied.

✔ A force concentrated over a small area will create a large pressure.

✔ A force spread out over a large area will create a small pressure.

✔ Pressure = Force / Area.

✔ Pressure in a liquid acts in all directions.

✔ Pressure in a liquid increases with depth.

✔ Liquids are incompressible and therefore can be used to transmit pressures and forces. Systems which do this are called hydraulic systems e.g. a hydraulic jack and hydraulic brakes.

✔ Air particles around us create a pressure called atmospheric pressure.

✔ We measure atmospheric pressure using a barometer.

End-of-Chapter Questions

1 Explain in your own words the following key terms you have met in this chapter:

pressure	hydraulic jack
pascal	force multiplier
water pressure	hydraulic brake
incompressible	air pressure
transmit	barometer
hydraulics	

2 Explain the following:

a) a sharp knife will cut through a piece of meat more easily than a blunt knife

b) a footballer wears boots that have studs on the soles

c) a pupil sits on a four-legged chair. When he rocks back so that he is sitting on just two legs, he damages the surface of the wooden floor.

3 The diagram below shows a crate which weighs 120 N and measures 3 m by 4 m by 5 m.

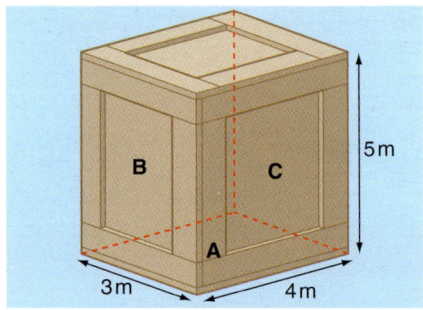

Calculate the pressure exerted on the floor by the crate when it is standing on each of the faces A, B and C.

4 **a)** Explain in detail what happens to a spherical balloon which is taken deeper and deeper below the surface of the sea by a diver.

b) Explain what happens to the balloon when the diver releases it and it rises to the surface.

c) Find out what divers mean when they talk about the bends. Why does it happen and how is it avoided?

5 **a)** Calculate the pressure transmitted through the oil in the hydraulic jack drawn below.

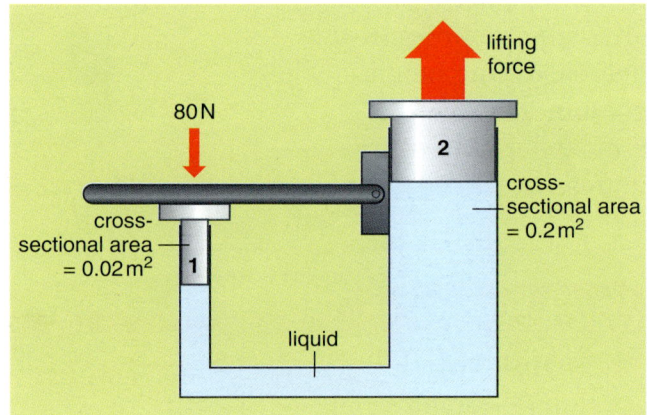

b) Calculate the force applied to piston 2.

c) Explain why a hydraulic jack will not work if there are air bubbles in the oil.

14 Moments

Figure 1 shows examples of forces which we might apply to objects in order to make them turn or rotate. This turning effect of a force is called a **moment**. The size of a moment depends upon

- the size of the force we apply
- the place where we apply the force and its direction.

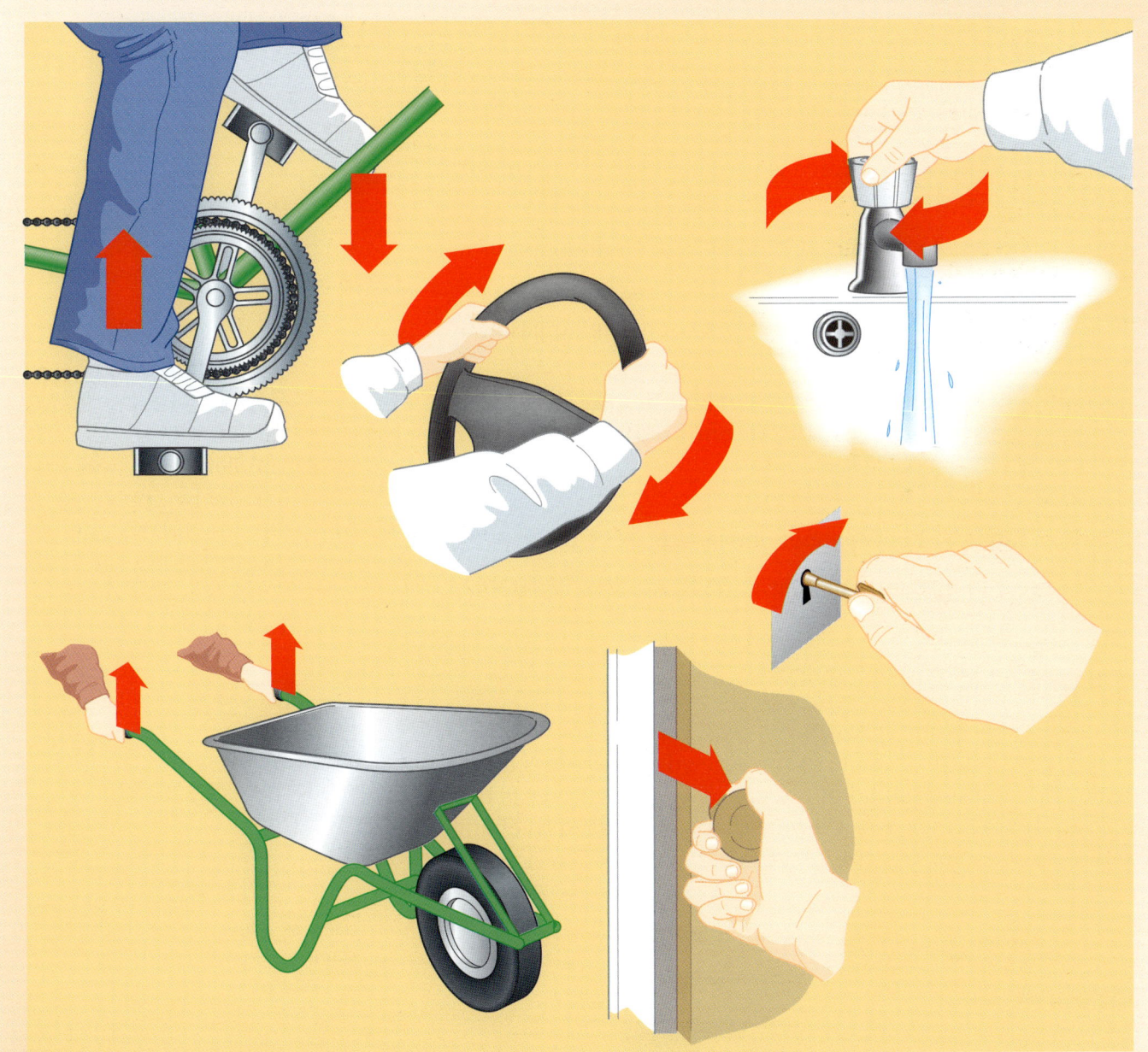

Figure 1 ▲ Turning forces

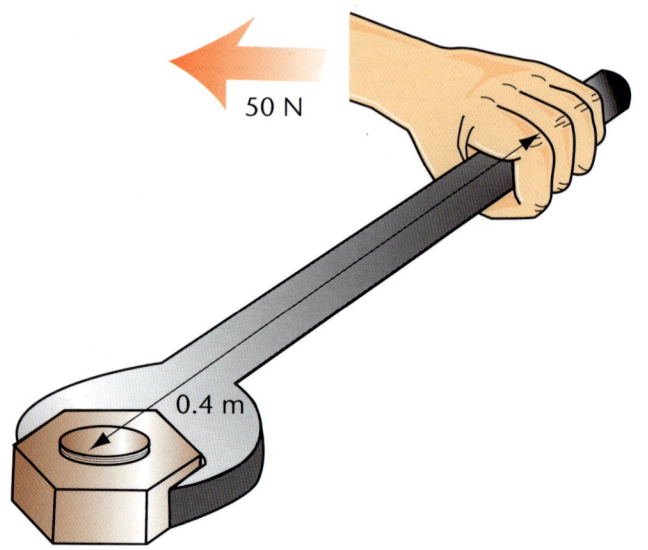

Figure 2 ▲ If we apply a force a long way from the pivot, we can create a large moment

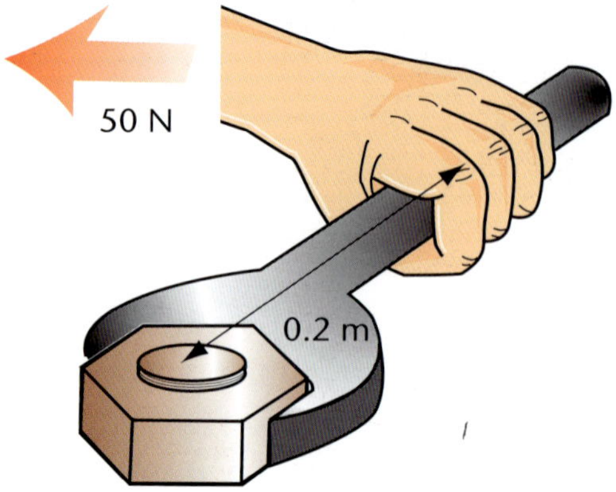

Figure 3 ▲ If we apply a similar force close to the pivot, the moment created will be smaller

The point around which a force is turning is called the **pivot**, sometimes called the **fulcrum**.

Figures 2 and 3 help explain why it is easier to undo a stiff nut using a long spanner.

The size of a moment can be calculated using the equation

moment of a force = force × perpendicular distance from pivot

Example

Moment created by long spanner above = 50 N × 0.4 m = 20 Nm

Moment created by short spanner = 50 N × 0.2 m = 10 Nm

Test Yourself

1 What is a moment?

2 Give three everyday examples of forces creating moments.

3 What is a pivot?

4 Calculate the moment created when a gardener lifts the handles of a wheel barrow by applying a force of 100 N. The distance from the handles to the pivot is 1.25 m.

Levers

Opening a tin of paint or syrup is easy when you know how. Pushing down on the handle of the screwdriver creates a large upward force lifting the lid of the can. The screwdriver is being used as a **lever**. A lever is a device that can be used to change the direction and/or the size of a force.

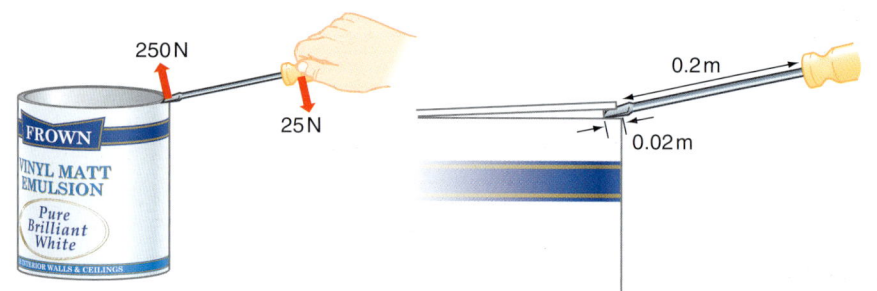

Figure 4 ◄ The screwdriver is used as a lever to open the tin of paint

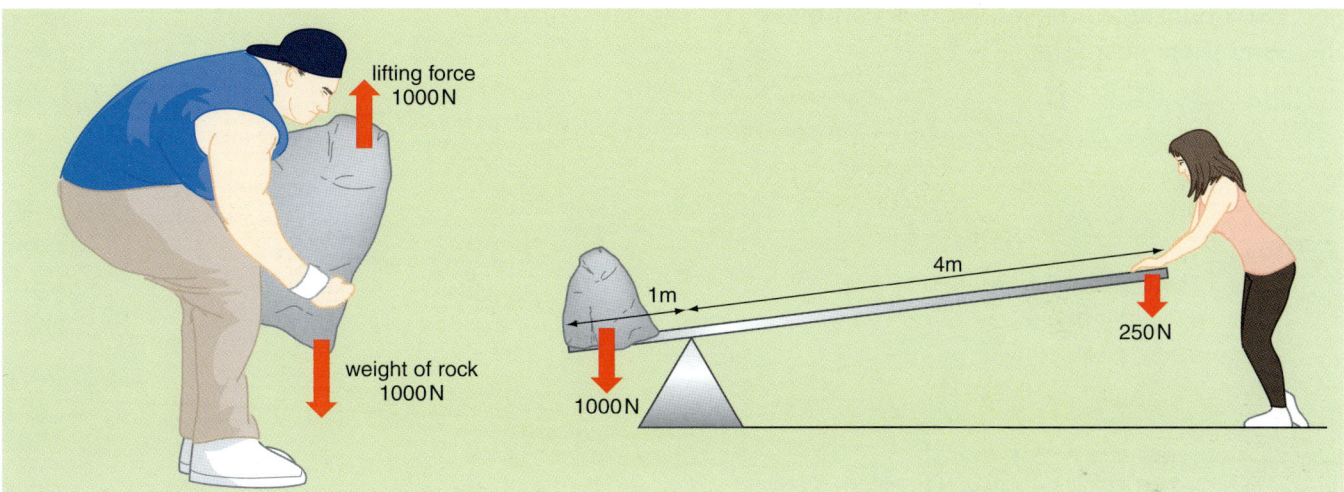

Figure 5 ▲

When we push down on the handle of the screwdriver, we create a moment. In this case the moment is

moment = force × distance from pivot = 25 N × 0.2 m = 5 Nm

As the lid begins to move, the moments on both sides of the pivot are almost equal.

Therefore, 5 Nm = Force × 0.02 m

Force = 5 Nm / 0.02 = 250 N

We can see from the above that by using the screwdriver as a lever we have increased the size of the force applied to the lid. A lever used in this way is called a **force multiplier**.

Lifting the rock in Figure 5 without a lever would be very difficult. With a lever, it is much easier.

Force needed to lift rock without lever is 1000 N

Force needed to lift rock using the lever is 250 N

Balancing moments

In Indian wrestling you try to create a larger moment than your opponent so that you can push his hand down onto the table. If the match is evenly **balanced**, there will be no movement, as the moment trying to turn the arm clockwise is equal to the moment trying to turn it anticlockwise.

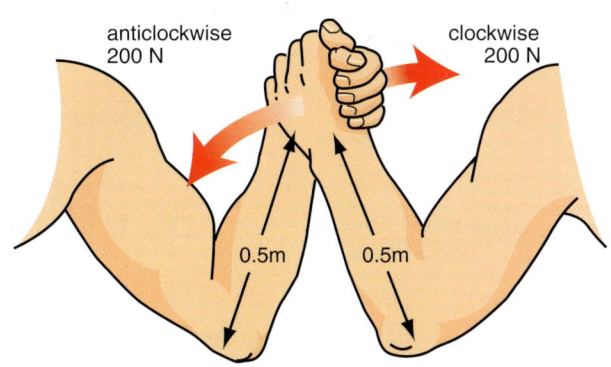

Figure 6 ▶   Balanced moments

Science Scope: PHYSICS

The two children in these pictures can make the see-saw turn by changing their positions.

Figure 7 ◄

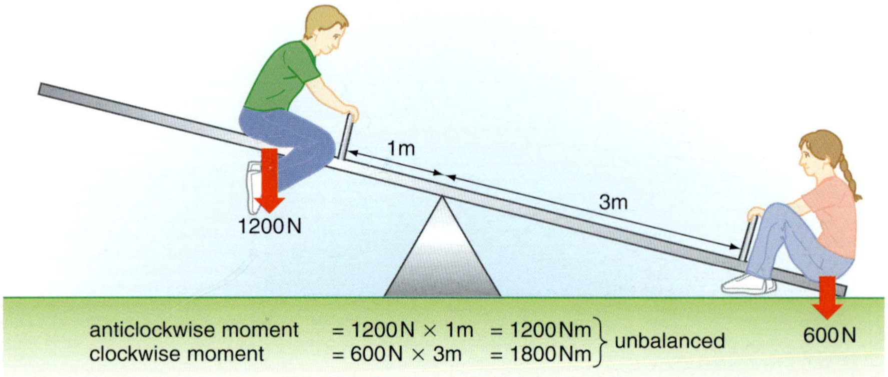

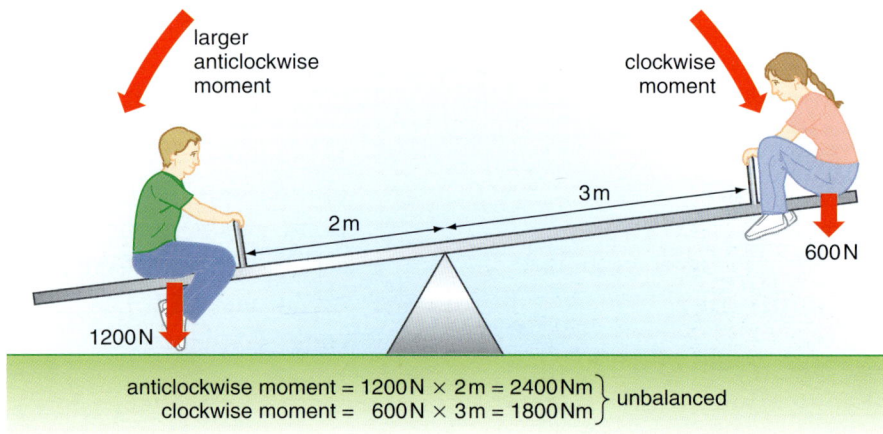

Sitting in the positions in Figure 7, the clockwise moment created by the girl is larger than the anticlockwise moment created by the boy. The see-saw therefore turns clockwise.

Figure 8 ◄

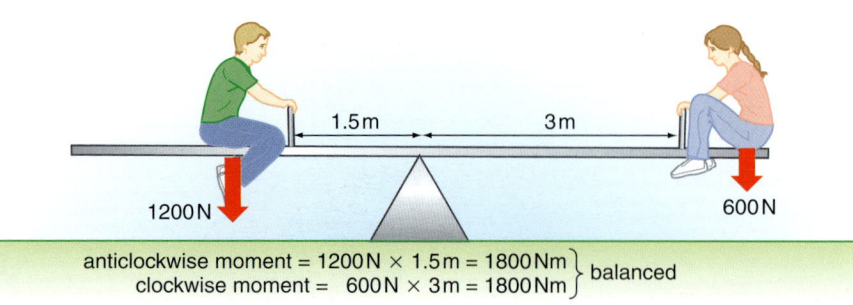

Sitting in the positions in Figure 8, the anticlockwise moment created by the boy is larger than the clockwise moment created by the girl. The see-saw therefore turns anticlockwise.

Figure 9 ◄

Sitting in the positions in Figure 9, the moments created by the children are equal. The see-saw is balanced.

7 Give two examples of balanced moments.

8 Which of these planks will a) tip and b) be balanced.

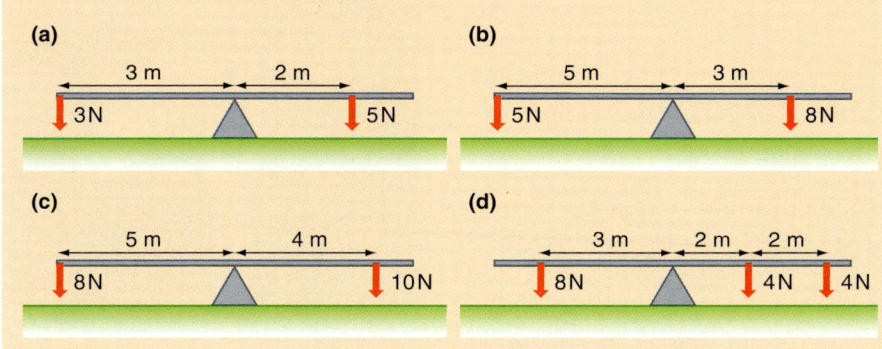

Centre of gravity (mass)

Sometimes it is useful to imagine that the weight of an object is concentrated at one place rather than spread throughout it. We call this place the **centre of gravity** or **centre of mass** of the object. For regular shaped objects the centre of gravity is often at the **geometric centre**.

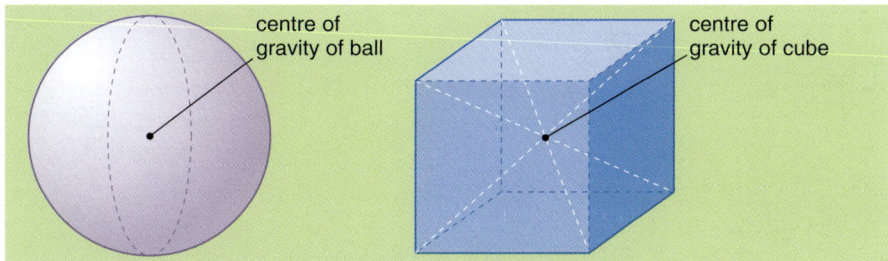

Figure 11 ◄

By imagining that all the weight of this ruler is concentrated at its centre, we can easily explain how we are able to balance it on one finger. All the weight of the ruler is directly above the pivot (the finger), so there are no clockwise or anticlockwise moments to cause the ruler to turn. If we move our finger to one side, the weight of the ruler creates a moment causing it to turn.

Figure 12 ▼ The ruler is only balanced when its centre is directly over the pivot

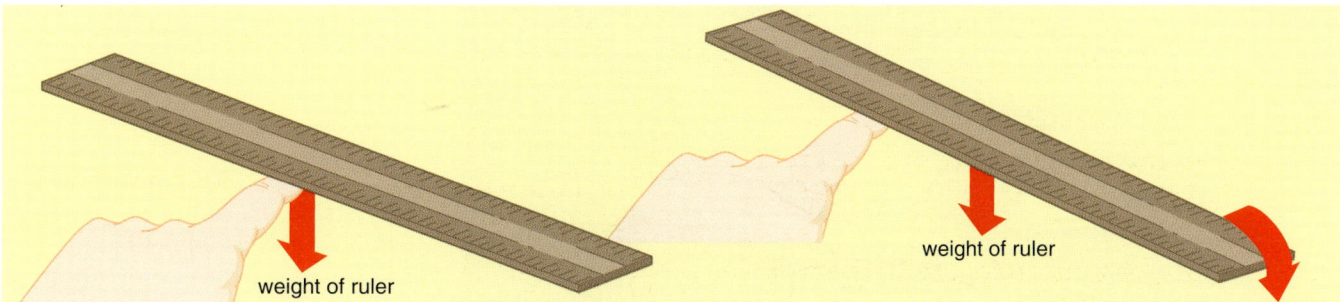

Science Scope: PHYSICS

The gymnast in Figure 13 must avoid creating any moments that will cause her to topple from the beam. To do this she must try to keep her centre of gravity directly above the beam. If the gymnast feels that she is losing her balance, she can try to move her centre of gravity back over the beam by altering the shape of her body i.e. holding out an arm or a leg.

If we were to stand on a beam or stand on just one foot, we might describe our situation as being **unstable**. Any small movement is likely to create a moment which would cause us to topple.

The glass in Figure 14a is also described as being unstable. When it is tilted by a small amount, its centre of gravity moves outside its base AB creating a clockwise moment.

The glass in Figure 14b is **stable**. Even after being tilted through a large angle, the weight of the glass is still acting through its base creating a **restoring moment** i.e. when released, the glass will return to its original position rather than toppling over.

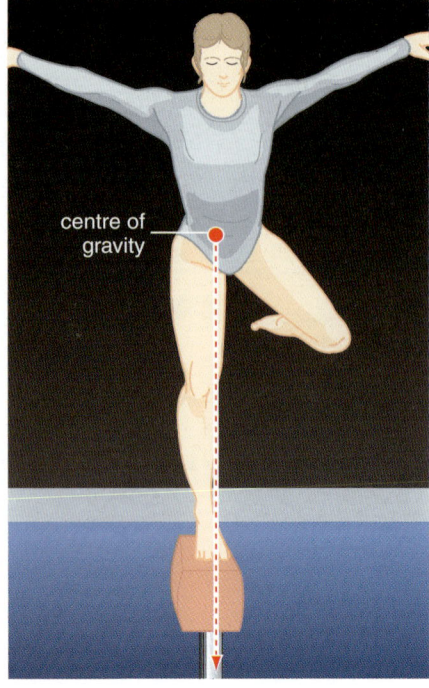

Figure 13 ▲ This gymnast is balanced

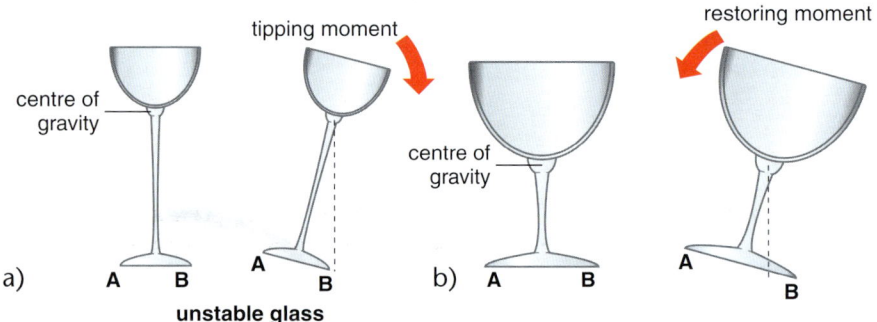

Figure 14 ◄ a) This glass is unstable. b) This glass is stable

As we can see from the above examples, if an object is to be stable it should be designed so that it has a low centre of gravity and a wide base.

Figure 15 ◄ This racing car is very stable. Its centre of gravity is very low and its wheel base is much wider than that of a normal car.

Test Yourself

9 Where is the centre of gravity of a ball?

10 Explain the difference between an object which is stable and one which is unstable.

11 Suggest two reasons why old fashioned racing cars at the beginning of the 20th Century were much more unstable than modern day racing cars.

Summary

When you have finished studying this chapter, you should understand that:

✔ Forces can cause objects to turn. This turning effect is called a moment.

✔ The moment of a force = size of force × perpendicular distance from pivot.

✔ A lever is used to change the size and/or the direction of an applied force.

✔ If an object is in equilibrium, i.e. does not turn, the clockwise moments must equal the anticlockwise moments.

✔ Stable objects are likely to have a low centre of gravity and a wide base.

End-of-Chapter Questions

1 Explain in your own words the following key terms you have met in this chapter:

moment   geometric centre

pivot (fulcrum) unstable object

lever   stable object

force multiplier restoring moment

balanced moments

centre of gravity (mass)

2 Given the apparatus shown below, devise an experiment to convince a friend that the turning effect of a force depends upon the size of the force and the perpendicular distance of the force from the pivot.

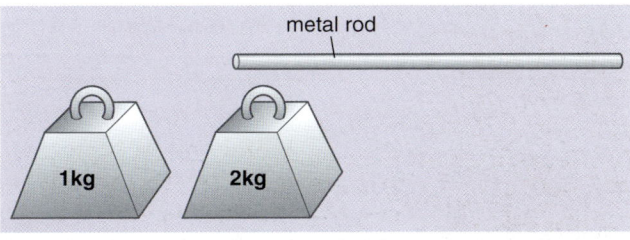

3 Explain why it is easier to open the door using handle A rather than handle B.

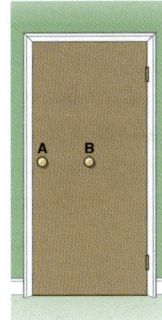

4 a) Calculate the moment created by the boy in the diagram below.

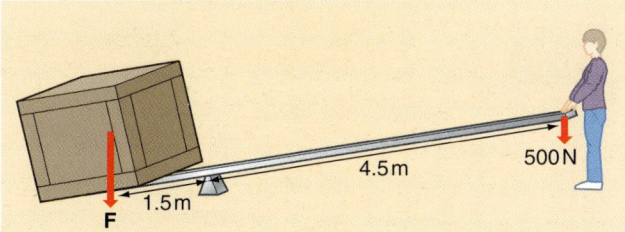

b) Calculate the weight of the crate if the boy is just able to lift it when applying a force of 500 N to the lever.

End-of-Chapter Questions continued

5 Calculate the weight of the girl if the see-saw drawn below is balanced.

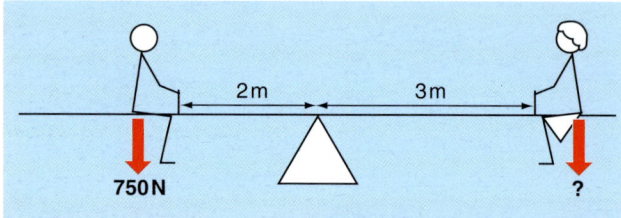

6 Why do tight rope walkers often carry long poles?

7 **a)** Explain why the rugby player on the right is in a more stable position.

b) A wire is stretched between two stands. Weights are placed on the bases of the stands to make them more stable. Explain why the weights make the stands more stable.

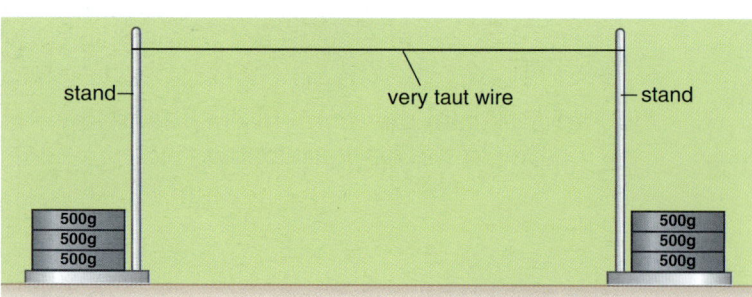

c) The diagram below shows a crane lifting a heavy weight. Why does the crane need a counter weight?

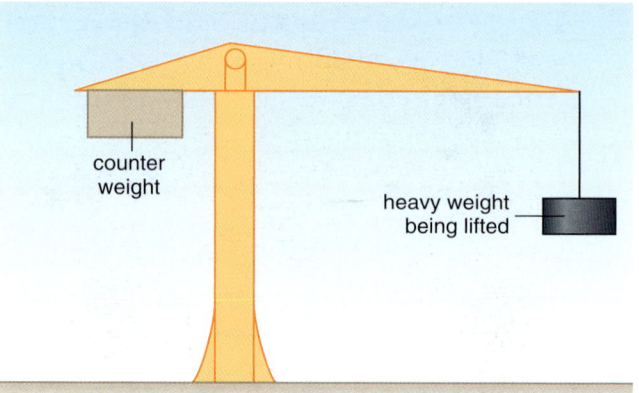

8 Your arm behaves as a lever which is pivoted at the elbow.

a) Find out how forces are applied to your 'lever'.

b) Draw a diagram and explain how your arm uses this lever to lift an object.

15

Speeding up

Figure 1 ▲

In order to put this space shuttle in orbit around the Earth, its motors must create enormous forces which will cause the space shuttle to **accelerate** to speeds of about 10 km/s. Once in orbit, however, the shuttle will continue at this speed without the need for any propulsive force. In this chapter we will be looking at how the speeds at which objects travel are affected by the forces being applied to them.

Speed

The **speed** of an object, tells us how far an object travels in a certain time.

Figure 2 ▲ These cars are travelling at around 110 km/h. They will travel 110 km each hour

Figure 3 ▲ This jogger is travelling at 5 m/s. She will travel 5 m each second

Table of speeds

	Typical speed
Walker	2 m/s
Sprinter	10 m/s
Cheetah (top speed)	28 m/s
Grand Prix racing car (top speed)	400 km/h
Sound waves in air	340 m/s
Light waves in air or vacuum	300 000 000 m/s

Table 1 ▲

It is possible using some instruments to measure directly the speed of an object. The policeman in the photograph below is using a 'radar gun' to obtain an instant measurement of the speed of a car.

Figure 4 ▲   This policeman is measuring the speed of passing cars

Without these kinds of instruments, we can still determine the speed of an object providing we have two pieces of information

- how far the object has travelled and
- how long it has taken to travel this distance.

We can then calculate the speed of the object using the equation

$$\text{Speed} = \text{Distance} / \text{Time}$$

Example 1

Calculate the speed of a cyclist who travels 1000 m in 50 s.

Speed = Distance / Time
Speed = 1000 m / 50 s
Speed = 20 m/s

Example 2

Calculate the speed of a train, which travels 500 km in 2.5 h.

Speed = Distance / Time
Speed = 500 km / 2.5 h
Speed = 200 km/h

It is highly unlikely that the cyclist or the train in the above examples travelled at a constant speed throughout their journey. The cyclist will be travelling much slower than 20 m/s as he starts. The train during its journey may have stopped several times at stations or at red signals. The speeds, which have been calculated, are **average speeds**.

Test Yourself

1   Is the speed measured by the policeman in Figure 4 an instantaneous speed or an average speed? Explain your answer.

2   Calculate the speed of a car that travels 300 km in 5 h.

3   Calculate the speed of a bird that flies 1500 m in 20 s.

Using the formula triangle

We can write the equation Speed (s) = Distance (d) / Time (t) as a **formula triangle**.

We can use the triangle to calculate any of the three quantities, speed, distance or time. We cover over with a finger the quantity a question is asking us to calculate and the triangle shows us how to calculate it.

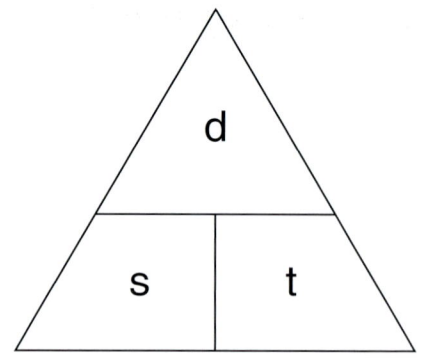

Figure 5 ▲ A formula triangle for the equation
speed = distance / time

Example 3

A bus travels at 80 km/h for 3 h. How far has the bus travelled?

The formula triangle shows us that

Distance = Speed × Time   (d = s × t)
Distance = 80 km/h × 3 h
Distance = 240 km

Example 4

How long will it take a jogger travelling at 5 m/s to travel 600 m?

The formula triangle shows us that

Time   = Distance / Speed $(t = \frac{d}{s})$
Time   = 600 m / 5 m/s
Time   = 120 s

Test Yourself

4   Calculate the missing values in the table below:

Speed	Distance	Time
A	200 m	5 s
B	1500 km	25 h
40 m/s	C	5 s
25 m/s	D	10 mins
60 km/h	420 km	E
35 m/s	7 km	F

Calculating the speed of sound

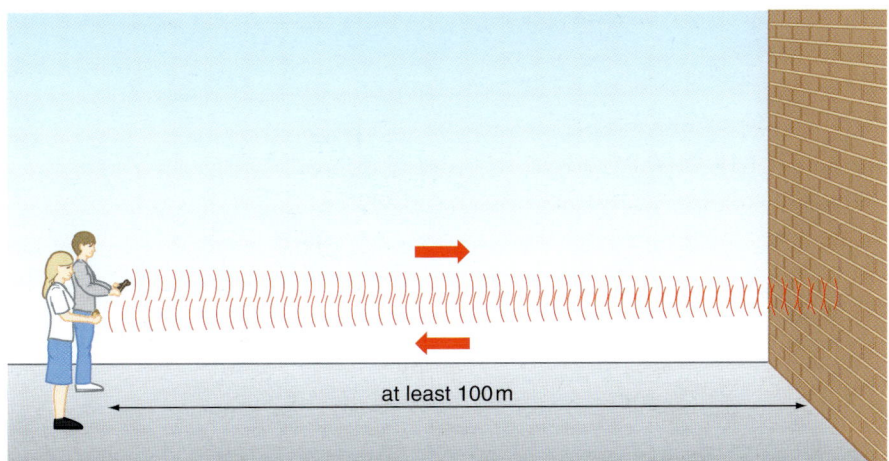

Figure 6 ▲　Calculating the speed of sound

Two pupils stand at least 100 m from a tall wall or building. One of the pupils bangs two pieces of wood together, so that they make a loud sound. The second pupil starts her stopwatch. She stops the watch when she hears an echo of the sound from the wall. The time on the watch is noted and the experiment repeated several times so that an average value for the time is calculated. The distance between the wall or building and the pupil is also measured.

Using the equation Speed = Distance travelled / Time taken, the speed of sound can be calculated.

Test Yourself

5　Using the results given below, calculate a value for the speed of sound.

Time before echo is heard:　1st reading　0.82 s
2nd reading　0.89 s
3rd reading　0.93 s

Distance to wall: 150 m

6　Explain why the value of the speed of sound calculated using the method described above is not affected by wind.

Graphs of motion

Sometimes it is useful to describe the journey of an object in the form of a graph.

Distance–time graphs

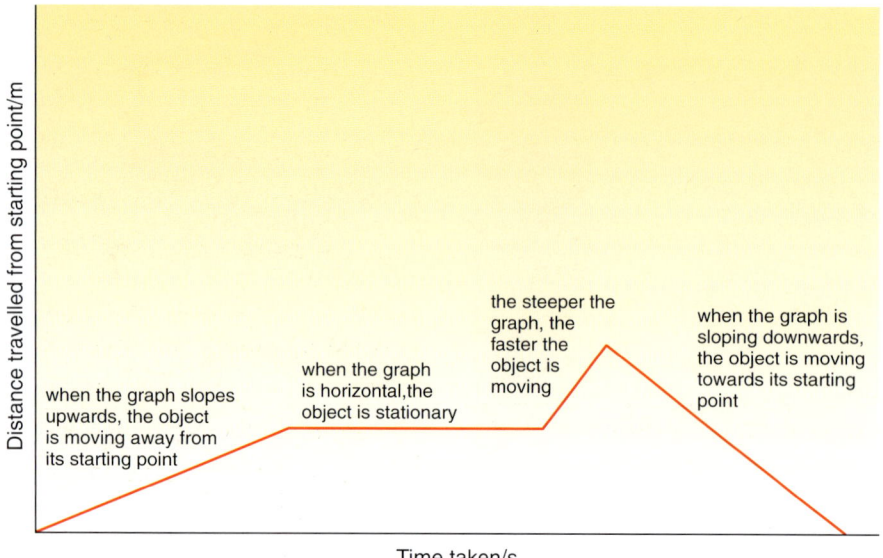

Figure 7 ▲ An example of a distance–time graph

A description of the above graph might be as follows. A man leaves his home and walks to his local shop. He stays in the shop for several minutes before hurrying further down the road to post a letter in a post box. Immediately after posting the letter, he jogs home.

Test Yourself

7   Sketch a distance–time graph for the following journey. (Assume that each part of the journey takes approximately the same time.)

A motorist travels slowly along busy roads and at a **constant speed**. When he reaches the motorway, he travels at a much higher constant speed. At the end of the motorway, there is a traffic jam and he is stationary for some time. Eventually he is able to continue his journey but at a very low speed.

The distance–time graph drawn on page 57 shows the journey of a cyclist. Because the axes now have values, it is possible to calculate the speed of the cyclist during his journey.

● Between A and B the cyclist is travelling away from his starting point. His speed during this part of the journey is equal to the distance travelled / time taken = x m / y s = z m/s
    $= \dfrac{200\,\text{m}}{40\,\text{s}} = 5\,\text{m/s}$

● Between B and C the cyclist is **stationary.**

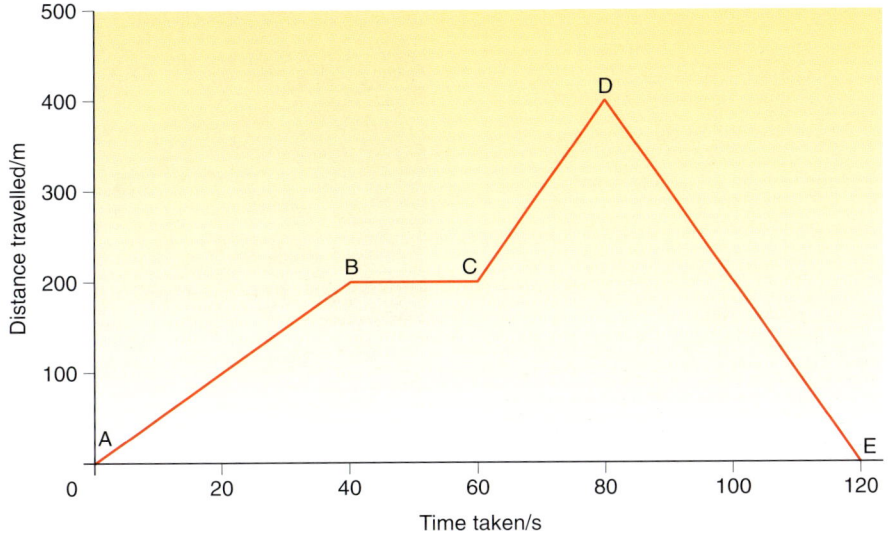

Figure 8 ▲ A distance–time graph for a cyclist

- Between C and D, the cyclist is moving away from his starting point. His speed during this part of his journey is equal to the distance travelled / time taken = x m / y s = z m/s

$$= \frac{200\,\text{m}}{20\,\text{s}} = 10\,\text{m/s}$$

- Between D and E, the cyclist is travelling back to his starting point. His speed during this part of his journey is equal to the distance travelled / time taken.

$$= \frac{400\,\text{m}}{40\,\text{s}} = 10\,\text{m/s}$$

How do forces affect speed?

As we have already seen in Chapter 2, applying a force to an object can affect its motion in several ways.

- A force can cause an object to speed up.
- A force can cause an object to slow down.
- A force can cause an object to change direction.

In Figures 9 and 10, the presence of forces and their effects on movements are used to our advantage. Sometimes these effects can be a disadvantage.

As objects travel through gases or liquids, they experience **frictional forces (drag)**. The effects of these **resistive forces** can be reduced if the object is **streamlined** (see page 25).

Figure 9 ▲   The frictional forces of the chute slow the car down

Figure 10 ▲   Astronauts use forces from their 'jet packs' to manoeuvre and change direction

Figure 11 ▲   The frictional forces on these dolphins are reduced because they are so perfectly streamlined in shape

Test Yourself

9   An object is moving through water. What will be the effect on the object's motion if
   a) a force is applied to the object, in the same direction as it is moving?
   b) a force is applied in the opposite direction to which it is moving?
   c) the object is given a more streamlined shape?

Speed–time graphs

A second kind of graph that could be used to describe the motion of an object is the speed–time graph.

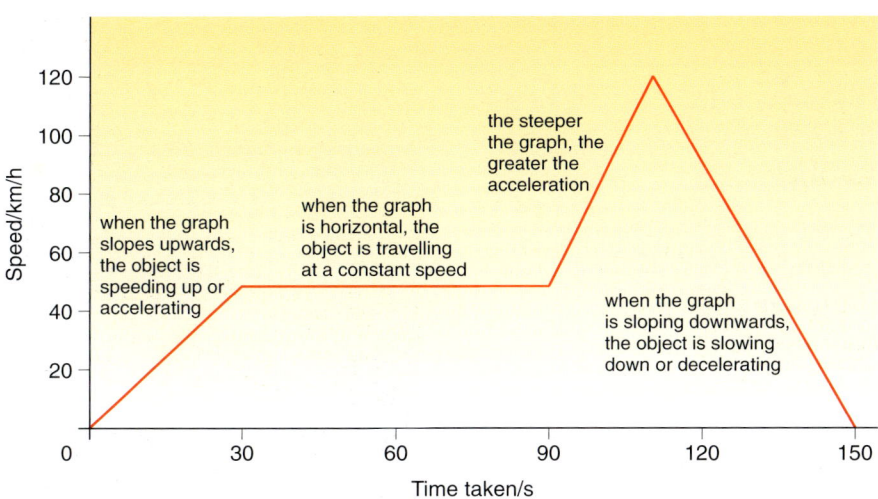

Figure 12 ◄   An example of a speed–time graph

A description of the above graph might be as follows. A driver starting from rest accelerates to a speed of 50 km/h in 30 s. He continues at this speed for a minute, before accelerating to 120 km/h in 20 s. Once he reaches this speed, he immediately **decelerates**, coming to a halt after 40 s.

Speed–time graph for a skydiver

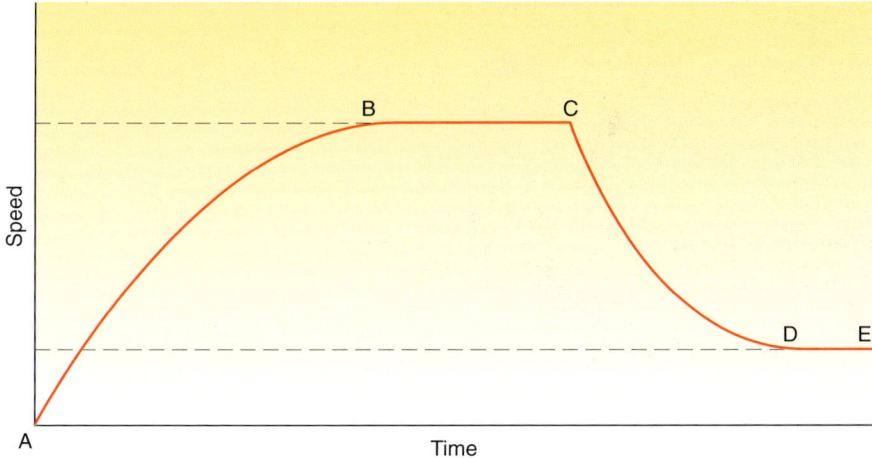

Figure 13 ◄   A speed–time graph for a skydiver

AB: As the skydiver jumps from her balloon, she accelerates. As her speed increases, the frictional forces of the air increase, causing her acceleration to decrease i.e. she does not speed up so quickly.

BC: Eventually the frictional forces of the air and the gravitational forces pulling her downwards balance. The skydiver now falls at a constant speed called her **terminal velocity**.

CD: When the skydiver opens her parachute, she increases the frictional forces of the air. Because these are acting upwards, she decelerates i.e. she slows down.

DE: As she slows, the frictional forces become less. Once again the upward forces and the downward forces become balanced and she falls at a much lower constant speed.

159

Extension box

If this large piece of card is moved as shown in the first diagram, it will cut through the air very easily and experience little resistance.

If the sheet is moved as shown in the second and third diagrams, large numbers of air particles are pushed closer together on one front side of the sheet creating a region of high pressure. Behind the sheet, the air particles are more spread out, creating a region of lower pressure. As a result of this pressure difference, there are large resistive forces (drag) exerted on the sheet. Usually the drag experienced by an object increases as its speed relative to the air increases.

If the sheet is tilted, as shown in the fourth diagram, air particles below and in front of the sheet will be pushed closer together, creating a region of higher pressure. Above and behind the sheet, the particles will be more spread out and so a region of lower pressure is created here. The result of this pressure difference is a force, part of which resists forward motion, part of which creates 'lift.'

In ski jumping, a competitor needs the surrounding air to give him lift whilst at the same time it must not provide too large a resistive force to his forward motion. Therefore, the angle at which he sets himself as he glides through the air is a compromise of these two requirements. If a competitor sets himself at the wrong angle, the consequences can be quite spectacular!

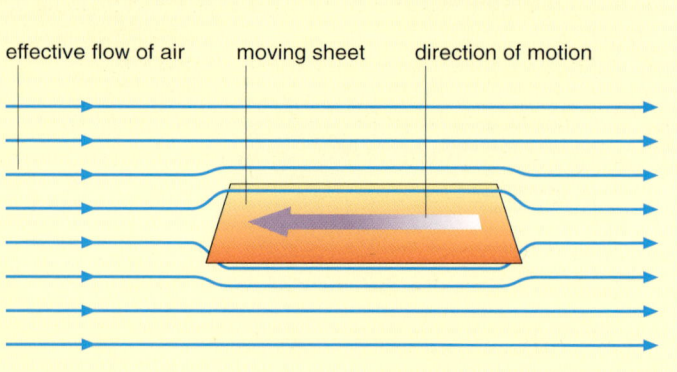

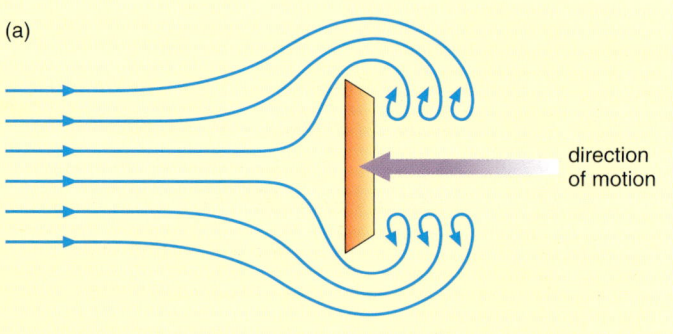

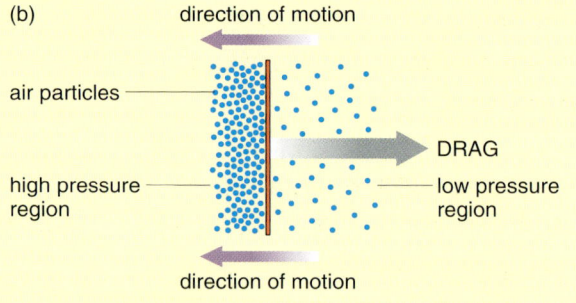

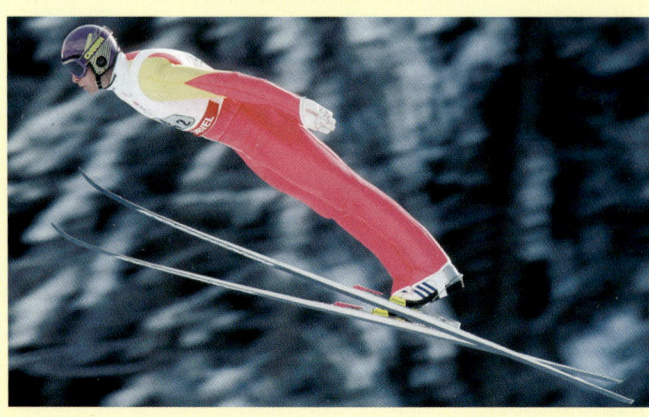

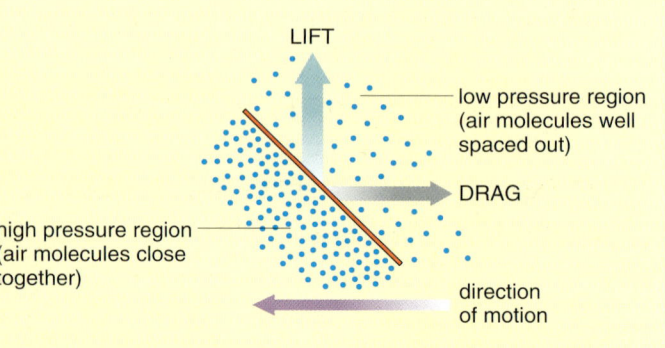

Summary

When you have finished studying this chapter, you should understand that:

✔ Speed = Distance travelled / Time taken

✔ If forces are applied to an object, they may change its motion. The greater the force, the greater the change.

✔ If the forces applied to an object are balanced, there is no change to its motion.

✔ Resistive forces can be reduced by streamlining.

✔ The faster an object travels, the greater the frictional forces it will experience.

✔ The motions of objects can be represented as distance–time graphs or speed–time graphs.

End-of-Chapter Questions

1 Explain in your own words the following key terms you have met in this chapter:

speed
accelerate
average speed
decelerate
constant speed
stationary

resistive forces
frictional forces
streamlined
drag
terminal velocity

2 A car travels 1500 m in 2 minutes. Calculate the speed of the car in a) m/s and b) km/h.

3 A jumbo jet completes its journey in 6 hours. For the first hour, it travels at a speed of 700 km/h. For the next 2 hours, it travels at a speed of 800 km/h. For the final part of its journey, it travels at 900 km/h.

a) Calculate the distance travelled by the jet in each of the three parts of its journey.

b) Calculate the average speed of the jet for this journey.

c) Explain why an aircraft is likely to burn more fuel the faster it flies.

4 Draw a distance–time graph for the following journey.

A woman leaves her home and walks 100 m in 200 s to the local bus stop. She waits 30 s for her bus. When the bus arrives she gets on board and travels 1500 m in the same direction in 90 s. She gets off the bus at the park and jogs home. This final part of her journey takes 8 min.

Calculate the woman's average speed for this journey.

5 Draw a speed–time graph for a skydiver from the moment he jumps from an aircraft to the moment he reaches the ground. Label the different parts of the graph. In your own words describe the forces exerted on the diver during his fall and describe the effects of these forces on his motion.

6 A light year is the distance light travels in one year. If the speed of light is 300 000 km/s, calculate the value of one light year in km.

Index

page numbers in italics refer to illustrations